Mathematics
with Applications
in Business and Social Sciences

Guided Notebook

HAWKES LEARNING

Editors:
Danielle C. Bess,
Adam Flaherty,
Marvin Glover,
Lisa Hinton

Project Manager:
Claudia Vance

Lead Developers:
Robert Alexander,
Doug Chappell

Creative Director:
Tee Jay Zajac

Designers:
Trudy Gove,
Kenneth Hanson,
Patrick Thompson,
Tee Jay Zajac

Cover Design:
Trudy Gove

Composition Assistance:
QSI (Pvt.) Ltd.

VP Research & Development: Marcel Prevuznak

Director of Content: Kara Roché

A division of Quant Systems, Inc.

546 Long Point Road
Mount Pleasant, SC 29464

Printed in the United States of America 🇺🇸

10 9 8 7 6 5 4 3 2 1

ISBN: 978-1-64277-122-0

Table of Contents

Strategies for Academic Success

CHAPTER 0

Fundamental Concepts of Algebra

CHAPTER 1

Equations and Inequalities in One Variable

Strategies for
Academic Success

Strategies for Academic Success 🎓

How to Read a Math Textbook

Reading a textbook is very different than reading a book for fun. You have to concentrate more on what you are reading because you will likely be tested on the content. Reading a math textbook requires a different approach than reading literature or history textbooks because the math textbook contains a lot of symbols and formulas in addition to words. Here are some tips to help you successfully read a math textbook.

Don't Skim 📖

When reading math textbooks, look at everything: titles, learning objectives, definitions, formulas, text in the margins, and any text that is highlighted, outlined, or in bold. Also pay close attention to any tables, figures, charts, and graphs.

Minimize Distractions

Reading a math textbook requires much more concentration than a novel by your favorite author, so pick a study environment with few distractions and a time when you are most attentive.

🚩 Start at the Beginning

Don't start in the middle of an assigned section. Math tends to build on previously learned concepts and you may miss an important concept or formula that is crucial to understanding the rest of the material in the section.

Highlight and Annotate

Put your book to good use and don't be afraid to add comments and highlighting. If you don't understand something in the text, reread it a couple of times. If it is still not clear, note the text with a question mark or some other notation so you can ask your instructor about it.

Go through Each Step of Each Example 📝

Make sure you understand each step of an example. If you don't understand something, mark it so you can ask about it in class. Sometimes math textbooks leave out intermediate steps to save space. Try working through the examples on your own, filling in any missing steps.

Take Notes *< This is important!*

Write down important definitions, symbols or notation, properties, formulas, theorems, and procedures. Review these daily as you do your homework and before taking quizzes and tests. Practice rewriting definitions in your own words so you understand them better.

Notes 9-25-17:

- The opposite of a negative integer is a positive integer.

- To add two integers with the same signs add their absolute values and use their common sign

💻 Use Available Resources

Many textbooks have companion websites to help you understand the content. These resources may contain videos that help explain more complex steps or concepts. Try searching the internet for additional explanations of topics you don't understand.

Read the Material Before Class

Try to read the material from your book before the instructor lectures on it. After the lecture, reread the section again to help you retain the information as you look over your class notes.

Understand the Mathematical Definitions $+ \times =$

Many terms used in everyday English have a different meaning when used in mathematics. Some examples include equivalent, similar, average, median, and product. Two equations can be equivalent to one another without being equal. An average can be computed mathematically in several ways. It is important to note these differences in meaning in your notebook along with important definitions and formulas.

Try Reading the Material Aloud

Reading aloud makes you focus on every word in the sentence. Leaving out a word in a sentence or math problem could give it a totally different meaning, so be sure to read the text carefully and reread, if necessary.

Questions

1. Explain how taking notes can help you understand new concepts and skills while reading a math textbook.

2. Think of two more tips for reading a math textbook.

Strategies for Academic Success 🎓

Tips for Success in a Math Course

Read Your Textbook/Workbook

One of the most important skills when taking a math class is knowing how to read a math textbook. Reading a section before class and then reading it again afterwards is an important strategy for success in a math course. If you don't have time to read the entire assigned section, you can get an overview by reading the introduction or summary and looking at section objectives, headings, and vocabulary terms.

Take Notes ✏️

Take notes in class using a method that works for you. There are many different note-taking strategies, such as the Cornell Method and Concept Mapping. You can try researching these and other methods to see if they might work better than your current note-taking system.

Review

While the information is fresh in your mind, read through your notes as soon as possible after class to make sure they are readable, write down any questions you have, and fill in any gaps. Mark any information that is incomplete so that you can get it from the textbook or your instructor later.

📁 Stay Organized

As you review your notes each day, be sure to label them using categories such as definition, theorem, formula, example, and procedure. Try highlighting each category with a different colored highlighter.

Use Study Aids

Use note cards to help you remember definitions, theorems, formulas, or procedures. Use the front of the card for the vocabulary term, theorem name, formula name, or procedure description. Write the definition, the theorem, the formula, or the procedure on the back of the card, along with a description in your own words.

Practice, Practice, Practice!

Math is like playing a sport. You can't improve your basketball skills if you don't practice—the same is true of math. Math can't be learned by only watching your instructor work through problems; you have to be actively involved in doing the math yourself. Work through the examples in the book, do some practice exercises at the end of the section or chapter, and keep up with homework assignments on a daily basis.

🖩 Do Your Homework

When doing homework, always allow plenty of time to finish it before it is due. Check your answers when possible to make sure they are correct. With word or application problems, always review your answer to see if it appears reasonable. Use the estimation techniques that you have learned to determine if your answer makes sense.

Understand, Don't Memorize

Don't try to memorize formulas or theorems without understanding them. Try describing or explaining them in your own words or look for patterns in formulas so you don't have to memorize them. For example, you don't need to memorize every perimeter formula if you understand that perimeter is equal to the sum of the lengths of the sides of the figure.

Study

Plan to study two to three hours outside of class for every hour spent in class. If math is your most difficult subject, then study while you are alert and fresh. Pick a study time when you will have the least interruptions or distractions so that you can concentrate.

🕐 Manage Your Time

Don't spend more than 10 to 15 minutes working on a single problem. If you can't figure out the answer, put it aside and work on another one. You may learn something from the next problem that will help you with the one you couldn't do. Mark the problems that you skip so that you can ask your instructor about it during the next class. It may also help to work a similar, but perhaps easier, problem.

Questions

1. Based on your schedule, what are the best times and places for you to study for this class?

2. Describe your method of taking notes. List two ways to improve your method.

Strategies for Academic Success 🎓

Tips for Improving Math Test Scores

Preparing for a Math Test

- Avoid cramming right before the test and don't wait until the night before to study. Review your notes and note cards every day in preparation for quizzes and tests.

- If the textbook has a chapter review or practice test after each chapter, work through the problems as practice for the test.

- If the textbook has accompanying software with review problems or practice tests, use it for review.

- Review and rework homework problems, especially the ones that you found difficult.

- If you are having trouble understanding certain concepts or solving any types of problems, schedule a meeting with your instructor or arrange for a tutoring session (if your college offers a tutoring service) well in advance of the next test.

Test-Taking Strategies

- Scan the test as soon as you get it to determine the number of questions, their levels of difficulty, and their point values so you can adequately gauge how much time you will have to spend on each question.

- Start with the questions that seem easiest or that you know how to work immediately. If there are problems with large point values, work them next since they count for a larger portion of your grade.

- Show all steps in your math work. This will make it quicker to check your answers later once you are finished since you will not have to work through all the steps again.

- If you are having difficulty remembering how to work a problem, skip it and come back to it later so that you don't spend all of your time on one problem.

After the Test

- The material learned in most math courses is cumulative, which means any concepts you miss on each test may be needed to understand concepts in future chapters. That's why it is extremely important to review your returned tests and correct any misunderstandings that may hinder your performance on future tests.

- Be sure to correct any work you did wrong on the test so that you know the correct way to do the problem in the future. If you are not sure what you did wrong, get help from a peer who scored well on the test or schedule time with your instructor to go over the test.

- Analyze the test questions to determine if the majority came from your class notes, homework problems, or the textbook. This will give you a better idea of how to spend your time studying for the next test.

- Analyze the errors you made on the test. Were they careless mistakes? Did you run out of time? Did you not understand the material well enough? Were you unsure of which method to use?

- Based on your analysis, determine what you should do differently before the next test and where you should focus your time.

Questions

1. Determine the resources that are available to you to help you prepare for tests, such as instructor office hours, tutoring center hours, and study groups.

2. Discuss two additional test taking strategies.

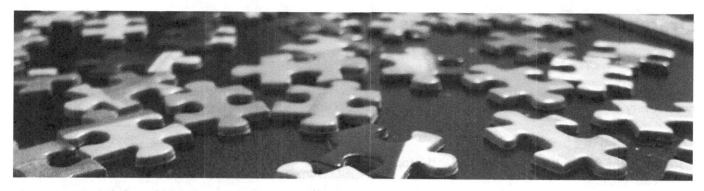

Strategies for Academic Success 🎓

Practice, Patience, and Persistence!

Have you ever heard the phrase "practice makes perfect"? This saying applies to many things in life. You won't become a concert pianist without many hours of practice. You won't become an NBA basketball star by sitting around and watching basketball on TV. The saying even applies to riding a bike. You can watch all of the videos and read all of the books on riding a bike, but you won't learn how to ride a bike without actually getting on the bike and trying to do it yourself. The same idea applies to math. Math is not a spectator sport.

Math is not learned by sleeping with your math book under your pillow at night and hoping for osmosis (a scientific term implying that math knowledge would move from a place of higher concentration—the math book—to a place of lower concentration—your brain). You also don't learn math by watching your professor do hundreds of math problems while you sit and watch. Math is learned by doing. Not just by doing one or two problems, but by doing many problems. Math is just like a sport in this sense. You become good at it by doing it, not by watching others do it. You can also think of learning math like learning to dance. A famous ballerina doesn't take a dance class or two and then end up dancing the lead in The Nutcracker. It takes years of practice, patience, and persistence to get that part.

Now, we aren't suggesting that you dedicate your life to doing math, but at this point in your education, you've already spent quite a few years studying the subject. You will continue to do math throughout college—and your life. To be able to financially support yourself and your family, you will have to find a job, earn a salary, and invest your money—all of which require some ability to do math. You may not think so right now, but math is one of the more useful subjects you will study.

It's important not only to practice math when taking a math course, but also to be patient and not expect immediate success. Just like a ballerina or NBA basketball star, who didn't become exceptional athletes overnight, it will take some time and patience to develop your math skills. Sure, you will make some mistakes along the way, but learn from those mistakes and move on.

Practice, patience, and persistence are especially important when working through applications or word problems. Most students don't like word problems and, therefore, avoid them. You won't become good at working word problems unless you practice them over and over again. You'll need to be patient when working through word problems in math since they will require more time to work than typical math skills exercises. The process of solving word problems is not a quick one and will take patience and persistence on your part to be successful.

Just as you work your body through physical exercise, you have to work your brain through mental exercise. Math is an excellent subject to provide the mental exercise needed to stimulate your brain. Your brain is flexible and it continues to grow throughout your life span—but only if provided the right stimuli. Studying mathematics and persistently working through tough math problems is one way to promote increased brain function. So, when doing mathematics, remember the 3 P's—Practice, Patience, and Persistence—and the positive effects they will have on your brain!

Questions

1. What is another area (not mentioned here) that requires practice, patience, and persistence to master? Can you think of anything you could master without practice?

2. Can you think of an example in your study of math where practice, patience, and persistence have helped you improve?

Strategies for Academic Success 🎓

Note Taking

Taking notes in class is an important step in understanding new material. While there are several methods for taking notes, every note-taking method can benefit from these general tips.

General Tips

- Write the date and the course name at the top of each page.
- Write the notes in your own words and paraphrase.
- Use abbreviations, such as ft for foot, # for number, def for definition, and RHS for right-hand side.
- Copy all figures or examples that are presented during the lecture.
- Review and rewrite your notes after class. Do this on the same day, if possible.

There are many different methods of note taking and it's always good to explore new methods. A good time to try out new note-taking methods is when you rewrite your class notes. Be sure to try each new method a few times before deciding which works best for you. Presented here are three note-taking methods you can try out. You may even find that a blend of several methods works best for you.

Note-Taking Methods

Outline

An outline consists of several topic headings, each followed by a series of indented bullet points that include subtopics, definitions, examples, and other details.

Example:

1. Ratio
   a. Comparison of two quantities by division.
   b. Ratio of a to b
      i. $\dfrac{a}{b}$
      ii. $a : b$
      iii. a to b
   c. Can be reduced
   d. Common units can cancel

Split Page

The split page method divides the page vertically into two columns with the left column narrower than the right column. Main topics go in the left column and detailed comments go in the right column. The bottom of the page is reserved for a short summary of the material covered.

Example:

Keywords:	Notes:
Ratios	1. Comparison of two quantities by division
	2. $\dfrac{a}{b}$, $a : b$, a to b
	3. Can reduce
	4. Common units can cancel

Summary: Ratios are used to compare quantities and units can cancel.

Mapping

The mapping method is the most visual of the three methods. One common way to create a mapping is to write the main idea or topic in the center and draw lines, from the main idea to smaller ideas or subtopics. Additional branches can be created from the subtopics until all of the key ideas and definitions are included. Using a different color for subtopic can help visually organize the topics.

Example:

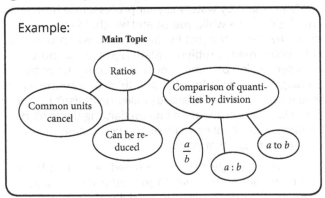

Questions

1. Find two other note taking methods and describe them.

2. Write five additional abbreviations that you could use while taking notes.

Strategies for Academic Success 🎓

Do I Need a Math Tutor?

If you do not understand the material being presented in class, if you are struggling with completing homework assignments, or if you are doing poorly on tests, then you may need to consider getting a tutor. In college, everyone needs help at some point in time. What's important is to recognize that you need help before it's too late and you end up having to retake the class.

Alternatives to Tutoring

Before getting a tutor, you might consider setting up a meeting with your instructor during their office hours to get help. Unfortunately, you may find that your instructor's office hours don't coincide with your schedule or don't provide enough time for one-on-one help.

Another alternative is to put together a study group of classmates from your math class. Working in groups and explaining your work to others can be very beneficial to your understanding of mathematics. Study groups work best if there are three to six members. Having too many people in a study group may make it difficult to schedule a time for all group members to meet. A large study group may also increase distractions. If you have too few people and those that attend are just as lost as you, then you aren't going to be helpful to each other.

Where to Find a Tutor

Many schools have both group and individual tutoring available. In most cases, the cost of this tutoring is included in tuition costs. If your college offers tutoring through a learning lab or tutoring center, then you should take advantage of it. You may need to complete an application to be considered for tutoring, so be sure to get the necessary paperwork at the start of each semester to increase your chances of getting a tutoring time that works well with your schedule. This is especially important if you know that you struggle with math or haven't taken any math classes in a while.

If you find that you need more help than the tutoring center can provide, or your school doesn't offer tutoring, you can hire a private tutor. The hourly cost to hire a private tutor varies significantly depending on the area you live in along with the education and experience level of the tutor. You might be able to find a tutor by asking your instructor for references or by asking friends who have taken higher-level math classes than you have. You can also try researching the internet for local reputable tutoring organizations in your area.

What to Look for in a Tutor

Whether you obtain a tutor through your college or hire a personal tutor, look for someone who has experience, educational qualifications, and who is friendly and easy to work with. If you find that the tutor's personality or learning style isn't similar to yours, then you should look for a different tutor that matches your style. It may take some effort to find a tutor who works well with you.

How to Prepare for a Tutoring Session

To get the most out of your tutoring session, come prepared by bringing your text, class notes, and any homework or questions you need help with. If you know ahead of time what you will be working on, communicate this to the tutor so they can also come prepared. You should attempt the homework prior to the session and write notes or questions for the tutor. Do not use the tutor to do your homework for you. The tutor will explain to you how to do the work and let you work some problems on your own while he or she observes. Ask the tutor to explain the steps aloud while working through a problem. Be sure to do the same so that the tutor can correct any mistakes in your reasoning. Take notes during your tutoring session and ask the tutor if he or she has any additional resources such as websites, videos, or handouts that may help you.

Questions

1. It's important to find a tutor whose learning style is similar to yours. What are some ways that learning styles can be different?

2. What sort of tutoring services does your school offer?

Strategies for Academic Success 🎓

Tips for Improving Your Memory

Experts believe that there are three ways that we store memories: first in the sensory stage, then in short term memory, and finally in long term memory.[1] Because we can't retain all the information that bombards us daily, the different stages of memory act as a filter. Your sensory memory lasts only a fraction of a second and holds your perception of a visual image, a sound, or a touch. The sensation then moves to your short term memory, which has the limited capacity to hold about seven items for no more than 20 to 30 seconds at a time. Important information is gradually transferred to long term memory. The more the information is repeated or used, the greater the chance that it will end up in long term memory. Unlike sensory and short term memory, long term memory can store unlimited amounts of information indefinitely. Here are some tips to improve your chances of moving important information to long-term memory.

Be attentive and focused on the information.

Study in a location that is free of distractions and avoid watching TV or listening to music with lyrics while studying.

Recite information aloud.

Ask yourself questions about the material to see if you can recall important facts and details. Pretend you are teaching or explaining the material to someone else. This will help you put the information into your own words.

Associate the information with something you already know.

Think about how you can make the information personally meaningful—how does it relate to your life, your experiences, and your current knowledge? If you can link new information to memories already stored, you create "mental hooks" that help you recall the information. For example, when trying to remember the formula for slope using rise and run, remember that rise would come alphabetically before run, so rise will be in the numerator in the slope fraction and run will be in the denominator.

Use visual images like diagrams, charts, and pictures.

You can make your own pictures and diagrams to help you recall important definitions, theorems, or concepts.

Split larger pieces of information into smaller "chunks."

This is useful when remembering strings of numbers, such as social security numbers and telephone numbers. Instead of remembering a sequence of digits such as 555777213 you can break it into chunks such as 555 777 213.

Group long lists of information into categories that make sense.

For example, instead of remembering all the properties of real numbers individually, try grouping them into shorter lists by operation, such as addition and multiplication.

Use mnemonics or memory techniques to help remember important concepts and facts.

A mnemonic that is commonly used to remember the order of operations is "Please Excuse My Dear Aunt Sally," which uses the first letter of the words Parentheses, Exponents, Multiplication, Division, Addition, and Subtraction to help you remember the correct order to perform basic arithmetic calculations. To make the mnemonic more personal and possibly more memorable, make up one of your own.

Use acronyms to help remember important concepts or procedures.

An acronym is a type of mnemonic device which is a word made up by taking the first letter from each word that you want to remember and making a new word from the letters. For example, the word HOMES is often used to remember the five Great Lakes in North America where each letter in the word represents the first letter of one of the lakes: Huron, Ontario, Michigan, Erie, and Superior.

Questions

1. Create an original mnemonic or acronym for any math topic covered so far in this course.

2. Explain two ways you can incorporate these tips into your study routine.

---

1   Source: http://science.howstuffworks.com/life/inside-the-mind/human-brain/human-memory2.htm

Strategies for Academic Success 🎓

Overcoming Anxiety

People who are anxious about math are often just not good at taking math tests. If you understand the math you are learning but don't do well on math tests, you may be in the same situation. If there are other subject areas in which you also perform poorly on tests, then you may be experiencing test anxiety.

How to Reduce Math Anxiety

- Learn effective math study skills. Sit near the front of your class and take notes. Ask questions when you don't understand the material. Review your notes after class and read new material before it's covered in class. Keep up with your assignments and do a lot of practice problems.

- Don't accept negative self talk such as "I am not good at math" or "I just don't get it and never will." Maintain a positive attitude and set small math achievement goals to keep you positively moving toward bigger goals.

- Visualize yourself doing well in math, whether it's on a quiz or test, or passing a math class. Rehearse how you will feel and perform on an upcoming math test. It may also help to visualize how you will celebrate your success after doing well on the test.

- Form a math study group. Working with others may help you feel more relaxed about math in general and you may find that other people have the same fears.

- If you panic or freeze during a math test, try to work around the panic by finding something on the math test that you can do. Once you gain confidence, work through other problems you know how to do. Then, try completing the harder problems, knowing that you have a large part of the test completed already.

- If you have trouble remembering important concepts during tests, do what is called a "brain drain" and write down all the formulas and important facts that you have studied on your test or scratch paper as soon as you are given the test. Do this before you look at any questions on the test. Having this information available to you should help boost your confidence and reduce your anxiety. Doing practice brain drains while studying can help you remember the concepts when the test time comes.

How to Reduce Test Anxiety

- Be prepared. Knowing you have prepared well will make you more confident and less anxious.

- Get plenty of sleep the night before a big test and be sure to eat nutritious meals on the day of the test. It's helpful to exercise regularly and establish a set routine for test days. For example, your routine might include eating your favorite food, putting on your lucky shirt, and packing a special treat for after the test.

- Talk to your instructor about your anxiety. Your instructor may be able to make accommodations for you when taking tests that may make you feel more relaxed, such as extra time or a more calming testing place.

- Learn how to manage your anxiety by taking deep, slow breaths and thinking about places or people who make you happy and peaceful.

- When you receive a low score on a test, take time to analyze the reasons why you performed poorly. Did you prepare enough? Did you study the right material? Did you get enough rest the night before? Resolve to change those things that may have negatively affected your performance in the past before the next test.

- Learn effective test taking strategies. See the study skill on Tips for Improving Math Test Scores.

> ### Questions
> 1. Describe your routine for test days. Think of two ways you can improve your routine to reduce stress and anxiety.
>
> 2. Research and describe the accommodations that your instructor or school can provide for test taking.

Strategies for Academic Success 🎓

Online Resources

With the invention of the internet, there are numerous resources available to students who need help with mathematics. Here are some quality online resources that we recommend.

HawkesTV

tv.hawkeslearning.com

If you are looking for instructional videos on a particular topic, then start with HawkesTV. There are hundreds of videos that can be found by looking under a particular math subject area such as introductory algebra, precalculus, or statistics. You can also find videos on study skills.

YouTube

www.youtube.com

You can also find math instructional videos on YouTube, but you have to search for videos by topic or key words. You may have to use various combinations of key words to find the particular topic you are looking for. Keep in mind that the quality of the videos varies considerably depending on who produces them.

Google Hangouts

plus.google.com/hangouts

You can organize a virtual study group of up to 10 people using Google Hangouts. This is a terrific tool when schedules are hectic and it avoids everyone having to travel to a central location. You do have to set up a Google+ profile to use Hangouts. In addition to video chat, the group members can share documents using Google Docs. This is a great tool for group projects!

Wolfram|Alpha

www.wolframalpha.com

Wolfram|Alpha is a computational knowledge engine developed by Wolfram Research that answers questions posed to it by computing the answer from "curated data." Typical search engines search all of the data on the Internet based on the key words given and then provide a list of documents or web pages that might contain relevant information. The data used by Wolfram|Alpha is said to be "curated" because someone has to verify its integrity before it can be added to the database, therefore ensuring that the data is of high quality. Users can submit questions and request calculations or graphs by typing their request into a text field. Wolfram|Alpha then computes the answers and related graphics from data gathered from both academic and commercial websites such as the CIA's World Factbook, the United States Geological Survey, financial data from Dow Jones, etc. Wolfram|Alpha uses the basic features of Mathematica, which is a computational toolkit designed earlier by Wolfram Research that includes computer algebra, symbol and number computation, graphics, and statistical capabilities.

Questions

1. Describe a situation where you think Wolfram|Alpha might be more helpful than YouTube, and vice versa.

2. What are some pros and cons to using Google Hangouts?

Strategies for Academic Success 🎓

Preparing for a Final Math Exam

Since math concepts build on one another, a final exam in math is not one you can study for in a night or even a day or two. To pull all the concepts together for the semester, you should plan to start one or two weeks ahead of time. Being comfortable with the material is key to going into the exam with confidence and lowering your anxiety.

Before You Start Preparing for the Exam

1. What is the date, time, and location of the exam? Check your syllabus for the final exam time and location. If it's not on your syllabus, your instructor should announce this information in class.

2. Is there a time limit on the exam? If you experience test anxiety on timed tests, be sure to speak to your professor about it and see if you can receive accommodations that will help reduce your anxiety, such as extended time or an alternate testing location.

3. Will you be able to use a formula sheet, calculator, and/or scrap paper on the exam? If you are not allowed to use a formula sheet, you should write down important formulas and memorize them. Most of the time, math professors will advise you of the formulas you need to know for an exam. If you cannot use a calculator on the exam, be sure to practice doing calculations by hand when you are preparing for the exam and go back and check them using the calculator.

A Week Before the Exam

1. Decide where to study for the exam and with whom. Make sure it's a comfortable study environment with few outside distractions. If you are studying with others, make sure the group is small and that the people in the group are motivated to study and do well on the exam. Plan to have snacks and water with you for energy and to avoid having to delay studying to go get something to eat or drink. Be sure and take small breaks every hour or two to keep focused and minimize frustration.

2. Organize your class notes and any flash cards with vocabulary, formulas, and theorems. If you haven't used flash cards for vocabulary, go back through your notes and highlight the vocabulary. Create a formula sheet to use on the exam, if the professor allows. If not, then you can use the formula sheet to memorize the formulas that will be on the exam.

3. Start studying for the exam. Studying a week before the exam gives you time to ask your instructor questions as you go over the material. Don't spend a lot of time reviewing material you already know. Go over the most difficult material or material that you don't understand so you can ask questions about it. Be sure to review old exams and work through any questions you missed.

3 Days Before the Exam

1. Make yourself a practice test consisting of the problem types. Don't necessarily put the questions in the order that the professor covered them in class.

2. Ask your instructor or classmates any questions that you have about the practice test so that you have time to go back and review the material you are having difficulty with.

The Night Before the Exam

1. Make sure you have all the supplies you will need to take the exam: formula sheet and calculator, if allowed, scratch paper, plain and colored pencils, highlighter, erasers, graph paper, extra batteries, etc.

2. If you won't be allowed to use your formula sheet, review it to make sure you know all the formulas. Right before going to bed, review your notes and study materials, but do not stay up all night to "cram."

3. Go to bed early and get a good night's sleep. You will do better if you are rested and alert.

The Day of the Exam

1. Get up with plenty of time to get to your exam without rushing. Eat a good breakfast and don't drink too much caffeine, which can make you anxious.

2. Review your notes, flash cards, and formula sheet again, if you have time.

3. Get to class early so you can be organized and mentally prepared.

Checklist for the Exam

Date of the Exam: _____   **Time of the Exam:** _____

Location of the Exam: _____

Items to bring to the exam:

___ calculator and extra batteries

___ formula sheet

___ scratch paper

___ graph paper

___ pencils

___ eraser

___ colored pencils or highlighter

___ ruler or straightedge

Notes or other things to remember for exam day:

During the Exam

1. Put your name at the top of your exam immediately. If you are not allowed to use a formula sheet, before you even look at the exam, do what is called a "brain drain" or "data dump." Recall as much of the information on your formula sheet as you possibly can and write it either on the scratch paper or in the exam margins if scratch paper is not allowed. You have now transferred over everything on your "mental cheat sheet" to the exam to help yourself as you work through the exam.

2. Read the directions carefully as you go through the exam and make sure you have answered the questions being asked. Also, check your solutions as you go. If you do any work on scratch paper, write down the number of the problem on the paper and highlight or circle your answer. This will save you time when you review the exam. The instructor may also give you partial credit for showing your work. (Don't forget to attach your scratch work to your exam when you turn it in.)

3. Skim the questions on the exam, marking the ones you know how to do immediately. These are the problems you will do first. Also note any questions that have a higher point value. You should try to work these next or be sure to leave yourself plenty of time to do them later.

4. If you get to a problem you don't know how to do, skip it and come back after you finish all the ones you know how to do. A problem you do later may jog your memory on how to do the problem you skipped.

5. For multiple choice questions, be sure to work the problem first before looking at the answer choices. If your answer is not one of the choices, then review your math work. You can also try starting with the answer choices and working backwards to see if any of them work in the problem. If this doesn't work, see if you can eliminate any of the answer choices and make an educated guess from the remaining ones. Mark the problem to come back to later when you review the exam.

6. Once you have an answer for all the problems, review the entire exam. Try working the problems differently and comparing the results or substituting the answers into the equation to verify they are correct. Do not worry about finishing early. You are in control of your own time—and your own success!

Questions

1. Does your syllabus provide any of the information needed for the checklist?

2. Are there any tips or suggestions mentioned here that you haven't thought of before?

Strategies for Academic Success 🎓

Managing Your Time Effectively

Have you ever made it to the end of a day and wondered where all of your time went? Sometimes it feels like there aren't enough hours in the day. Managing your time is important because you can never get that time back. Once it's gone, you have to rush and cram the work into your schedule. Not only will you start feeling stressed out, but you may also find yourself turning in late or incomplete work.

Here are three strategies for managing your time more effectively.

🕐 Time Budgets

Time budgets help you find the time you need to complete necessary projects and tasks. Just like a financial budget shows you how you spend your money, a time budget shows you how you spend your time. You can then identify "wasted" time that could be used more productively.

To begin budgeting your time, assess how much time each week you spend on different types of activities, like Sleep, Meals, Work, Class, Study, Extracurricular, Exercise, Personal, Other, etc.

- What are some activities you'd like to spend more time doing in the future?

- What are some activities you should spend less time doing in the future?

Based on your answers to the questions above, create a weekly time budget. One week contains only 168 hours. If you want to spend more time on a particular activity, you'll need to find that time somewhere. Use a planner to schedule specific blocks of time for study sessions, meals, travel times, and morning/evening routines. As a general rule, you should set aside at least two hours of study time for every one hour of class time. That means that a three-credit course would require at least six hours of outside work per week.

⚖️ Breaks

When you are working on an important project or studying for a big exam, you can feel tempted to go as long as possible without taking a break. While staying focused is important, working yourself until you're mentally drained will lower the quality of your work and force you to take even more time recovering.

Just like taking breaks helps your physical body recover, it will also help your brain re-energize and refocus. During study sessions, you should plan to take a break at least once an hour. Study and work breaks should usually last around five minutes. The longer the break, the harder it is to start working again. Some courses have a built-in break during the middle of the class period. Stand up and move around, even if you don't feel tired. Even this little bit of physical movement can help you think more clearly.

📖 Avoiding Multitasking

Multitasking is working on more than one task at a time. When you have several assignments that need to be completed, you may be tempted to save time by working on two or three of them at once. While this strategy might seem like a time-saver, you will probably end up using more time than if you had done each task individually. Not only will you have to switch your focus from one task to the next, but you will also make more mistakes that will need to be corrected later. Multitasking usually ends up wasting time instead of saving it.

Instead of trying to do two things at once, schedule yourself time to work on one task at a time. To-do lists can be helpful tools for keeping yourself focused on finishing one item before moving on to another. You'll do better work and save yourself time.

Questions

1. Are there any areas in your day that are taking up too much of your time, making it hard to devote enough time to more important things?

2. Can you think of a time when multitasking has resulted in lower quality outcome in your experience?

Fundamental Concepts of Algebra

0.1 Real Numbers

Types of Real Numbers

The Natural (or Counting) Numbers: This is the set of numbers
_____. The set is infinite, so in list form we can write only the first few numbers.

The Whole Numbers: This is the set of natural numbers and 0: _____ .
Again, we can list only the first few members of this set. No special symbol will be assigned to this set in this text.

The Integers: This is the set of natural numbers, their negatives, and 0. As a list, this is the set
_____ . Note that the list continues indefinitely in both directions.

The Rational Numbers: This is the set, with symbol _____ (for quotient), of _____
_____ (hence the name). That is, any rational number can be written in the form $\frac{p}{q}$, where p and q are both integers and _____ . When written in decimal form, rational numbers either
_____ or have a _____ of digits past some point.

The Irrational Numbers: Every real number that is not _____ is, by definition, irrational. In decimal form, irrational numbers are _____ and _____ . No special symbol will be assigned to this set in this text.

The Real Numbers: Every set above is a _____ of the set of real numbers, which is denoted
_____ . Every real number is either _____ or _____ , and no real number is both.

Inequality Symbols (Order)

Symbol	Reading	Meaning
$a < b$	"a is _____ b"	a lies to the _____ of b on the number line.
$a \leq b$	"a is _____ b"	a lies to the _____ of b or is _____ b.
$b > a$	"b is _____ a"	b lies to the _____ of a on the number line.
$b \geq a$	"b is _____ a"	b lies to the _____ of a or is _____ ____ a.

The two symbols < and > are called _____ inequality signs, while the symbols $\leq$ and $\geq$ are _____ inequality signs.

CAUTION!

Remember that order is defined by the _____ of real numbers on the number line, *not* by _____ (its distance from zero). For instance, $-36 < 5$ because -36 lies to the _____ of 5 on the number line. Also, be aware that the negation of the statement $a \leq b$ is the statement _____ . Furthermore, if $a \leq b$ and $a \geq b$, then it must be the case that _____ .

Set-Builder Notation

The notation $\{x \mid x \text{ has property } P\}$ is used to describe a _____ , all of which have the _____ . This can be read "the _____ of all real numbers _____ having property _____ ."

The symbol "|" is also read as _____ so the above notation can also be read "the set of _____ , *such that* _____ ."

The Empty Set

A set with no elements is called the _____ or the _____, and is denoted by the symbol ∅.

The empty set can arise from a set defined using set-builder notation; for example, the set $\{y|y > 1 \text{ and } y \leq -4\}$ is equivalent to the _____, since no real number y satisfies the stated property.

Interval Notation

Interval Notation	Set-Builder Notation	Meaning
(a, b)	$\{x|a < x < b\}$	all real numbers _____ a and b
$[a, b]$	$\{x|a \leq x \leq b\}$	all real numbers _____ a and b, _____ both a and b
$(a, b]$	$\{x|a < x \leq b\}$	all real numbers _____ a and b, _____ b but _____ a
$(-\infty, b)$	$\{x|x < b\}$	all real numbers _____ b
$[a, \infty)$	$\{x|x \geq a\}$	all real numbers _____ a

Intervals of the form (a, b) are called _____ intervals, while those of the form $[a, b]$ are _____ intervals. The interval $(a, b]$ is _____ (or _____. Of course, a half-open interval may be open at either endpoint, as long as it is closed at the other. The symbols $-\infty$ and ∞ indicate that the interval extends _____ in the left and the right directions, respectively.

Note that $(-\infty, b)$ _____ the endpoint b, while $[a, \infty)$ _____ the endpoint a.

CAUTION!

The symbols $-\infty$ and ∞ are just that: _____! They are not real numbers, so they cannot be solutions to a given equation. The fact that they are symbols, and not numbers, means that they can _____ _____ in a set of real numbers. For this reason, a _____ always appears next to either $-\infty$ or ∞; a bracket should _____ appear next to either infinity symbol.

Absolute Value

The _____ of a real number a, denoted as $|a|$, is defined by:

$$|a| = \begin{cases} \rule{6cm}{0.4pt} & \text{if } a \geq 0 \\ \rule{6cm}{0.4pt} & \text{if } a < 0 \end{cases}$$

The absolute value of a number is also referred to as its _____ ; it is the nonnegative number corresponding to its _____ from the _____ . Note that 0 is the only real number whose absolute value is 0.

Distance on the Real Number Line

Given two real numbers a and b, the _____ between them is defined to be _____ . In particular, the distance between a and 0 is _____ or just _____ .

Properties of Absolute Value

For all real numbers a and b:

1. $|a| \geq 0$ (The _____ of a number is never _____ .)

2. $|-a| =$ _____

3. $a \leq$ _____

4. $|ab| =$ _____

5. $\left| \dfrac{a}{b} \right| = \dfrac{|a|}{|b|},$ _____

6. $|a + b| \leq |a| + |b|$ (This is called the _____ , as it is a reflection of the fact that one side of a triangle is never longer than the _____ of the other two sides.)

0.1 Exercises

1. Identify the following number. Choose all that apply.

   ○ Natural Number   ○ Whole Number   ○ Integer   ○ Rational Number

   ○ Irrational Number   ○ Real Number   ○ Undefined

   -2

2. Identify the following number. Choose all that apply.

   ○ Natural Number   ○ Whole Number   ○ Integer   ○ Rational Number

   ○ Irrational Number   ○ Real Number   ○ Undefined

   $\dfrac{1}{4}$

3. Identify the following number. Choose all that apply.

   ○ Natural Number   ○ Whole Number   ○ Integer   ○ Rational Number

   ○ Irrational Number   ○ Real Number   ○ Undefined

   2π

4. Plot the set $\{-1.5, 0, 1.5, 3.0\}$ on the number line.

5. Select **all** the symbols from the set $\{>, \geq, \leq, <\}$ that can be placed in the blank to make the following a true statement.

   $\dfrac{-1}{6}$ ____ $\dfrac{-2}{3}$

6. Write the following statement as an inequality:

12.0 is strictly greater than −19.0

7. Write the following statement as an inequality:

−11 is less than or equal to 2

8. Write the following as an interval using interval notation.

$\{x | 13 < x\}$

9. Write the following set as an interval using interval notation.

$\{x | -11 < x \le -6\}$

10. Evaluate the following expression.

$-|13 - 2|$

0.2 The Arithmetic of Algebraic Expressions

EXAMPLE 2: Evaluating Algebraic Expressions

Evaluate the following algebraic expressions.

a. $5x^3 - 16$ for $x = 4$

b. $-3x^2 - 2(x + y)$ for $x = -2$ and $y = 3$

Solution

Field Properties

In this table, a, b, and c represent arbitrary real numbers. The first five properties apply to addition and multiplication, while the last combines the two.

Name of Property	Additive Version	Multiplicative Version
Closure	$a + b$ is a _____	ab is a _____
Commutative	$a + b = $ _____	$ab = $ _____
Associative	$a + (b + c) = $ _____	$a(bc) = $ _____
Identity	$a + 0 = $ _____ $= a$	$a \cdot 1 = $ _____ $= a$
Inverse	$a + (-a) = $ _____	$a \cdot \dfrac{1}{a} = $ _____ (for $a \neq 0$)
Distributive	$a(b + c) = $ _____	

Cancellation Properties

Let A, B, and C be algebraic expressions.

Additive Cancellation: Adding the same quantity to _____ of an equation results in an equivalent equation.

$$\text{If } A = B, \text{ then } A + \underline{\quad\quad} = B + \underline{\quad\quad}.$$

Multiplicative Cancellation: Multiplying both sides of an equation by the same _____ _____ results in an equivalent equation.

$$\text{If } A = B \text{ and } C \neq 0, \text{ then } A \cdot \underline{\quad\quad} = B \cdot \underline{\quad\quad}.$$

Zero-Factor Property

Let A and B represent algebraic expressions. If the product of A and B is 0, then at least one of _____ and _____ is itself 0.

$$AB = 0 \text{ implies that } \underline{\quad\quad} \text{ or } \underline{\quad\quad} \text{ (or both).}$$

Order of Operations

Step 1: If the expression is a fraction, simplify the _____ **and** _____ **individually, according to the guidelines in the following steps.**

Step 2: Parentheses, braces and brackets are all used as _____ **symbols. Simplify expressions within each set of** _____ **symbols, if any are present, working from the innermost outward.**

Step 3: Simplify all _____ **(exponents) and** _____ **.**

Step 4: Perform all _____ **and** _____ **in the expression in the order they occur, working from** _____ **.**

Step 5: Perform all _____ **and** _____ **in the expression in the order they occur, working from** _____ **.**

EXAMPLE 6: Order of Operations

Simplify the following expressions using the correct order of operations.

a. $4 - \dfrac{6}{2}$

b. -3^2

c. $\dfrac{3 - \left(3\sqrt{4} - 2^2\right)\left(\dfrac{6}{-2}\right)}{3 + 2\,(-2)}$

Solution

Union

In this definition, A and B denote two sets, and are represented in the Venn diagram by circles. The operation of union is depicted in the diagram by shading.

The **union** of A and B, denoted _____ , is the set $\{x\,|\,x \in A$ _____ $x \in B\}$. That is, an element x is in $A \cup B$ if it is in the set _____ , the set _____ , or _____ . Note that the union of A and B contains both individual sets.

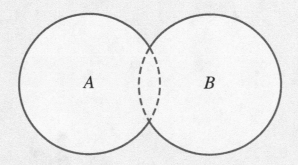

Intersection

In this definition, A and B denote two sets, and are represented in the Venn diagram by circles. The operation of intersection is depicted in the diagram by shading.

The **intersection** of A and B, denoted _____, is the set $\{x \mid x \in A$ _____ $x \in B\}$. That is, an element x is in $A \cap B$ if it is in _____ A and B. Note that the intersection of A and B is contained in each individual set.

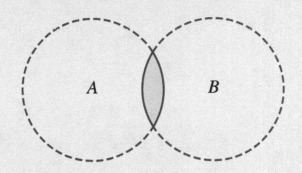

EXAMPLE 7: Union and Intersection of Intervals

Simplify the following unions and intersections of intervals.

a. $(-2, 4] \cup [0, 9]$

b. $(-2, 4] \cap [0, 9]$

c. $[3, 4) \cap (4, 9)$

d. $(-\infty, 4] \cup (-1, \infty)$

Solution

EXAMPLE 8: Union and Intersection

Simplify each of the following set expressions.

a. $\{1, 2\} \cup \{0, 3\}$

b. $\{x, y, z\} \cap \{w, x\}$

c. $\mathbb{Z} \cup \mathbb{R}$

d. $\mathbb{Z} \cap \mathbb{R}$

Solution

0.2 Exercises

1. Identify the terms in the following expression:

$$7x^5y^4 - (x - y) - 5xz$$

2. Evaluate the following expression for the given values of the variables: (Leave your answer in terms of π or use $\pi = 3.14$.)

$$-7x^3 + 7\pi y - y^2 \text{ for } x = 2 \text{ and } y = -1$$

3. Identify the property that justifies the following statement:

$$(3 - 3)\left(7^6\right) = \left(7^6\right)(3 - 3)$$

4. Identify the property that justifies the following statement:

$$5(-7y + 7) = -35y + 35$$

5. If the following statement is false, write False Statement. Otherwise, identify the property that justifies it. If one of the cancellation properties is being used to transform the equation, identify the quantity that is added to or multiplied by both sides.

$$7x - \frac{23}{8}y^6 - z = \frac{1}{8}y^6 - z \Leftrightarrow 7x - 3y^6 = 0$$

6. If the following statement is false, write False Statement. Otherwise, identify the property that justifies it. If one of the cancellation properties is being used to transform the equation, identify the quantity that is added to or multiplied by both sides.

$$\frac{-1}{8}x^8y = \frac{1}{4}(y + z) \Leftrightarrow \frac{-1}{3}x^8y = \frac{1}{3}(y + z)$$

7. Evaluate the following expression, expressing your answer in terms of π. Be sure to use the correct order of operations.

$$-7^2 + 8 \cdot \sqrt{5 + 2 \cdot 2} - 9\pi$$

8. Simplify the following union and/or intersection.

$[-5, 4) \cap (4, 20)$

9. Simplify the following union and/or intersection.

$(-9, 7] \cap [4, \infty)$

10. Simplify the following union and/or intersection.

$\mathbb{Z} \cap \mathbb{R}$

0.3 Integer Exponents

Natural Number Exponents

If a is any real number and if n is any natural number, then $a^n = \underbrace{a \cdot a \cdot \ldots \cdot a}_{n \text{ factors}}$. That is, a^n is just a shorter,

more precise way of denoting the product of _____.

In the expression a^n, a is called the _____, and n is the _____. The process of

multiplying n factors of a is called "raising a to the nth power," and the expression a^n may be referred to as

"the nth power of _____ or "a to the _____ power." Note that a^1 is simply a.

0 as an Exponent

For any real number $a \neq 0$, we define $a^0 = $_____. 0^0 is _____, just as division by 0 is

undefined.

Negative Integer Exponents

For any real number $a \neq 0$ and for any natural number n, $a^{-n} = $_____. (We don't allow a to be 0

simply to avoid the possibility of division by 0.) Since any negative integer is the negative of a natural

number, this defines exponentiation by negative integers.

Properties of Exponents

Throughout this table, a and b may be taken to represent constants, variables, or more complicated algebraic expressions. The letters n and m represent integers.

Property	Example
1. $a^n \cdot a^m = a^{n+m}$	$3^3 \cdot 3^{-1} = \underline{\hspace{1cm}} = 3^2 = 9$
2. $\dfrac{a^n}{a^m} = a^{n-m}$	$\dfrac{7^9}{7^{10}} = \underline{\hspace{1cm}} = 7^{-1}$
3. $a^{-n} = \dfrac{1}{a^n}$	$5^{-2} = \underline{\hspace{1cm}} = \dfrac{1}{25}$ and $x^3 = \underline{\hspace{1cm}}$
4. $(a^n)^m = a^{nm}$	$(2^3)^2 = \underline{\hspace{1cm}} = 2^6 = 64$
5. $(ab)^n = a^n b^n$	$(7x)^3 = \underline{\hspace{1cm}} = 343x^3$ and $(-2x^5)^2 = \underline{\hspace{1cm}} = 4x^{10}$
6. $\left(\dfrac{a}{b}\right)^n = \dfrac{a^n}{b^n}$	$\left(\dfrac{3}{x}\right)^2 = \underline{\hspace{1cm}} = \dfrac{9}{x^2}$ and $\left(\dfrac{1}{3z}\right)^2 = \underline{\hspace{1cm}} = \dfrac{1}{9z^2}$

Here we assume every expression is defined. That is, if an exponent is 0, then the base is nonzero, and if an expression appears in the denominator of a fraction, then that expression is nonzero. Remember that $a^0 = \underline{\hspace{1cm}}$ for every $a \neq 0$.

EXAMPLE 3: Properties of Exponents

Simplify the following expressions by using the properties of exponents. Write the final answers with only positive exponents. (As in the table of properties, it is assumed that every expression is defined.)

c. $\dfrac{\left(-2x^3 y^{-1}\right)^{-3}}{\left(18x^{-3}\right)^0 (xy)^{-2}}$

d. $\left(7xz^{-2}\right)^2 \left(5x^2 y\right)^{-1}$

Note:

There are often many ways to simplify an expression; the order in which you apply the properties of exponents will not change the result.

Solution

CAUTION!

Many errors can be made in applying the properties of exponents as a result of forgetting the exact form of the properties. The first column below contains examples of some common errors. The second column contains the corrected statements.

Incorrect	Correct
$x^2 x^5 = x^{10}$	$x^2 x^5 = x^{2+5} = \underline{}$
$2^4 2^3 = 4^7$	$2^4 2^3 = 2^{4+3} = \underline{}$
$(3+4)^2 = 3^2 + 4^2$	$(3+4)^2 = \underline{}$
$\left(x^2 + 3y\right)^{-1} = \dfrac{1}{x^2} + \dfrac{1}{3y}$	$\left(x^2 + 3y\right)^{-1} = \underline{}$
$(3x)^2 = 3x^2$	$(3x)^2 = 3^2 x^2 = \underline{}\ \underline{}$
$\dfrac{x^5}{x^{-2}} = x^3$	$\dfrac{x^5}{x^{-2}} = x^{5-(-2)} = \underline{}$

0.3 Exercises

1. Simplify the following expression.

$(-2)^6$

2. Simplify the following expression.

-5^6

3. Simplify the following expression.

$(-5)^3$

4. Simplify the following expression, writing your answer with only positive exponents.

$\dfrac{1}{8z^{-6}}$

5. Simplify the following expression, writing your answer with only positive exponents.

$$\frac{x^{-6}}{x^{-1}}$$

6. Simplify the following expression, writing your answer with only positive exponents.

$$\frac{n^3}{n}$$

7. Use the properties of exponents to simplify the following expression, writing your answer with only positive exponents.

$$\frac{y^9\left(y^{-2}z^{-2}\right)^{-2}}{\left(4y^4\right)^5 z^2}$$

8. Simplify the following expression, writing your answer with only positive exponents.

$$\left(x^{-2}\right)^{-4}$$

9. Simplify the following expression, writing your answer with only positive exponents.

$$\left(\frac{-4x^2}{y^{-2}}\right)^3$$

10. Simplify the following expression, writing your answer with only positive exponents.

$$\frac{\left(5x^2y^{-2}\right)^{-2}}{\left(4x^{-2}y\right)^{-3}}$$

0.4 Radicals

Radical Notation

Case 1: *n* is an even natural number. If a is a nonnegative real number and n is an even natural number, $\sqrt[n]{a}$ is the nonnegative real number b with the property that _____ . That is, $\sqrt[n]{a} = b$ if and only if $a = b^n$. Note that $\left(\sqrt[n]{a}\right)^n = a$ and $\sqrt[n]{a^n} = a$.

Case 2: *n* is an odd natural number. If a is any real number and n is an odd natural number, $\sqrt[n]{a}$ is the real number b (whose sign will be the _____ as the sign of a) with the property that $b^n = a$. Again, $\sqrt[n]{a} = b$ if and only if _____ , $\left(\sqrt[n]{a}\right)^n = a$, and $\sqrt[n]{a^n} = a$.

The expression $\sqrt[n]{a}$ gives the nth root of a in _____ . The natural number n is called the _____ , a is the _____ , and $\sqrt{}$ is called a _____ . By convention, $\sqrt[2]{}$ is usually simply written as $\sqrt{}$.

Perfect Powers

A **perfect square** is an integer equal to the _____ of another integer. The square root of a perfect square is always an _____ .

A **perfect cube** is an integer equal to the _____ of another integer. The cube root of a perfect cube is always an _____ .

EXAMPLE 2: The Pythagorean Theorem

Given a right triangle with sides of length a and b, the Pythagorean theorem states that the length of the hypotenuse c is given by $c = \sqrt{a^2 + b^2}$. In the Pythagorean theorem, find the following:

a. The radicand

b. The index

c. The value of c if $a = 5$ and $b = 12$

d. The value of c if $a = 1$ and $b = 2$

Solution

Simplified Form of Radical Expressions

A radical expression is in **simplified form** when:

1. The radicand contains no factor with an _____ greater than or equal to the index of the radical.

2. The radicand contains no _____ .

3. The denominator, if there is one, contains no _____ .

4. The greatest common factor of the index and any exponent occurring in the radicand is _____ . That is, the index and any exponent in the radicand have no _____ factor other than 1.

Properties of Radicals

Let a and b represent constants, variables, or more complicated algebraic expressions. The letters n and m represent natural numbers. Assume that all expressions are defined and are real numbers.

Property	Example
1. $\sqrt[n]{ab} = \sqrt[n]{a}\sqrt[n]{b}$	$\sqrt[3]{3x^6y^2} = \sqrt[3]{3} \cdot \sqrt[3]{x^3} \cdot \sqrt[3]{x^3} \cdot \sqrt[3]{y^2}$
	$= \sqrt[3]{3} \cdot \underline{\quad} \cdot \underline{\quad} \cdot \sqrt[3]{y^2} = x^2\sqrt[3]{3y^2}$
2. $\sqrt[n]{\dfrac{a}{b}} = \dfrac{\sqrt[n]{a}}{\sqrt[n]{b}}$	$\sqrt[4]{\dfrac{x^4}{16}} = \dfrac{\sqrt[4]{x^4}}{\sqrt[4]{16}} = \dfrac{\underline{\quad}}{2}$
3. $\sqrt[m]{\sqrt[n]{a}} = \sqrt[mn]{a}$	$\sqrt[3]{\sqrt{64}} = \sqrt[3]{\sqrt[2]{64}} = \underline{\quad} = 2$

EXAMPLE 3: Simplifying Radical Expressions

Simplify the following radical expressions:

a. $\sqrt[3]{-16x^8y^4}$

b. $\sqrt{8z^6}$

c. $\sqrt[3]{\dfrac{72x^2}{y^3}}$

Solution

CAUTION!

As with the properties of exponents, many mistakes arise from forgetting the properties of radicals. One common error is to rewrite $\sqrt{a+b}$ as $\sqrt{a} + \sqrt{b}$. These two expressions are _____ ! To convince yourself of this, evaluate the two expressions with actual constants in place of a and b. Using $a = 9$ and $b = 16$, we see that $5 = \sqrt{9+16} \neq \sqrt{9} + \sqrt{16} = 7$.

EXAMPLE 4: Rationalizing the Denominator

Simplify the following radical expressions:

a. $\dfrac{1}{\sqrt{x}}$

b. $\sqrt[5]{\dfrac{-4x^6}{8y^2}}$

c. $\dfrac{4}{\sqrt{7} + \sqrt{3}}$

d. $\dfrac{-\sqrt{5x}}{5 - \sqrt{x}}$

Note:

The first two examples follow Case 1. In general, it still helps to factor the radicands before further simplifying. The second pair of examples have two terms in the denominator, and thus follow Case 2. Begin by multiplying the numerator and denominator by the conjugate radical of the denominator.

Solution

0.4 Exercises

1. Determine if the following radical expression is a real number. If it is, you are to evaluate the expression.

$$\sqrt[3]{-125}$$

2. Determine if the following radical expression is a real number. If it is, you are to evaluate the expression.

$$\sqrt[3]{-\frac{27}{125}}$$

3. Simplify the following radical expression.

$$\sqrt[3]{-64y^{12}z^6}$$

4. Simplify the following radical expression.

$$\sqrt[4]{\frac{x^{28}z^{24}}{81}}$$

5. Simplify the following radical expression.

$$\sqrt{3xz^{10}}$$

6. Simplify the following radical by rationalizing the denominator.

$$\frac{\sqrt{z}+\sqrt{y}}{\sqrt{z}-\sqrt{y}}$$

7. Simplify the following radical by rationalizing the denominator.

$$\frac{\sqrt{x}}{\sqrt{x} - \sqrt{6}}$$

8. Rationalize the numerator of the following expression (that is, rewrite the expression as an equivalent expression that has no radicals in the numerator).

$$\frac{5 + \sqrt{z}}{6}$$

9. Simplify the following radical expression by rationalizing the denominator.

$$\sqrt{\frac{6z}{5x}}$$

10. Simplify the following radical expression by rationalizing the denominator.

$$\frac{\sqrt{90}}{\sqrt{14}}$$

0.5 Rational Exponents

EXAMPLE 1: Combining Radicals

Combine the radical expressions, if possible.

a. $-3\sqrt{8x^5} + \sqrt{18x}$

b. $\sqrt[3]{54x^3} + \sqrt{50x^2}$

c. $\sqrt{\dfrac{1}{12}} - \sqrt{\dfrac{25}{48}}$

Solution

Rational Number Exponents

Meaning of $a^{\frac{1}{n}}$: If n is a natural number and if $\sqrt[n]{a}$ is a real number, then $a^{\frac{1}{n}} = $ _____ .

Meaning of $a^{\frac{m}{n}}$: If m and n are natural numbers with $n \neq 0$, if m and n have no common factors greater than 1, and if $\sqrt[n]{a}$ is a real number, then $a^{\frac{m}{n}} = $ _____ $= $ _____ . Either $\sqrt[n]{a^m}$ or $\left(\sqrt[n]{a}\right)^m$ can be used to evaluate $a^{\frac{m}{n}}$, as they are equal. $a^{-\frac{m}{n}}$ is defined to be $\dfrac{1}{a^{\frac{m}{n}}}$.

EXAMPLE 2: Simplifying Expressions

Simplify each of the following expressions, writing your answer using the same notation as the original expression.

b. $\sqrt[9]{-8x^6}$

d. $\sqrt[5]{\sqrt[3]{x^2}}$

Solution

0.5 Exercises

1. Simplify the following expression.

$$\sqrt[5]{\sqrt[7]{y^{140}}}$$

2. Simplify the following expression.

$$\left(5y^4 + 6\right)^{\frac{2}{3}} \left(5y^4 + 6\right)^{\frac{-1}{3}}$$

3. Simplify the following expression.

$$64^{\frac{-4}{3}}$$

4. Simplify the following expression.

$$\sqrt[3]{z^7} \sqrt[9]{z^6}$$

5. Combine the radical expressions, if possible.

$$\sqrt[5]{-64z^6} + 8z\sqrt[5]{2z}$$

6. Combine the radical expressions, if possible.

$$\sqrt{3y} - \sqrt[4]{3y}$$

7. Combine the radical expressions, if possible.

$$\sqrt{8y^3} + 8y\sqrt{2y}$$

8. Convert the given rational expression to radical notation.

$$y^{\frac{9}{2}}$$

9. Convert the given radical expression to rational exponent notation.

$$\sqrt[7]{z^8}$$

10. Simplify the following expression.

$$\sqrt[3]{z^5} \cdot \sqrt[9]{z^6}$$

0.6 Polynomials and Factoring

Classification of Polynomials

		Example
Monomial	(polynomial with _____)	$5x$
Binomial	(polynomial with _____	$7x - 8$
Trinomial	(polynomial with _____	$x^3 + 9x - 4$
Polynomial	(any sum or difference of a set of monomials)	$7x - 8, x^3 + 9x - 4$

Example 3: Subtraction with Polynomials

Find the difference $\left(9x^3 - 4x^2 + 3x\right) - \left(8x^3 + 4x^2 - 5\right)$.

Solution

Caution

Be careful to note that the two trinomials $X^2 - AX + A^2$ and $X^2 + AX + A^2$ in Formulas IV and V are

not _____ trinomials.

Example 6: Using Special Products

Find and name each of the following products.

a. $9x^2 - 16$

   Solution

b. $(x + 5)^2$

   Solution

c. $(x+3)\left(x^2 - 3x + 9\right)$

Solution

Steps for Factoring Polynomials

1. Look for any _____ factors (usually monomials or binomials).

2. See if the product fits any of the special forms just listed.

3. Try the reverse of the _____ method if the product is a trinomial.

Example 8: Using the FOIL Method

Use the FOIL method to factor the following polynomials.

a. $x^2 + 8x + 12$

Solution

b. $2x^2 - 13x - 7$

Solution

Example 9: Using Special Products

Factor each of the following polynomials by using the list of special products.

a. $4x^2 - 25$

   Solution

b. $x^2 + 12x + 36$

   Solution

c. $x^3 - 125$

   Solution

0.6 Exercises

1. Multiply the polynomials using the distributive property and combine like terms.

$(x + 7)(x + 2)$

2. Multiply the polynomials using the distributive property and combine like terms.

$(3x + 8y)(x + 2y)$

3. Multiply the polynomials using the distributive property and combine like terms.

$(x - 4)(x^2 + x + 1)$

4. Multiply the polynomials using the distributive property and combine like terms.

$(x + 2)(x^2 - 2x + 4)$

5. Perform the indicated operation by removing the parentheses and combining like terms.

$$\left(-7x^2 + 2x\right) - \left(-8x^2 - 9x\right)$$

6. Perform the indicated operation by removing the parentheses and combining like terms.

$$\left(6x^2 - 8x - 4\right) + \left(7x^2 - 5\right)$$

7. Factor the given polynomial by finding the greatest common monomial factor (or the negative of the greatest common monomial factor) and rewrite the expression.

$$6xy - 14y^3$$

8. Factor the following trinomial.

$$y^2 + 10y + 21$$

9. Factor the following trinomial.

$$z^2 - 13z + 42$$

10. Factor the following trinomial.

$$6x^2 - 7x - 5$$

Equations and Inequalities in One Variable

1.1 Linear Equations in One Variable

Linear Equations in x

If a, b, and c are constants and $a \neq 0$, then a _____ **equation in x** is an equation that can be written in the form

$$ax + b = c.$$

Note: A linear equation in x is also called a **first-degree equation in x** because the variable x can be written with the exponent 1. That is, $x = x^1$.

Addition Property of Equality

If the same algebraic expression is added to both sides of an equation, the new equation has the same solutions as the original equation. Symbolically, if A, B, and C are algebraic expressions, then the equations

$$A = B$$

and

_____

have the same solutions. Equations with the same solutions are said to be **equivalent**.

Multiplication (or Division) Property of Equality

If both sides of an equation are multiplied by (or divided by) the same nonzero constant, the new equation has the same solutions as the original equation. Symbolically, if A and B are algebraic expressions and C is any nonzero constant, then the equations

$$A = B$$

and _____ _____   where $C \neq 0$

and _____ _____   where $C \neq 0$

have the same solutions and are **equivalent**.

Solving Linear Equations

1. Simplify each side of the equation by removing any grouping symbols and combining like terms. (In some cases, you may want to multiply both sides of the equation by a constant to clear fractional or decimal coefficients.)

2. Use the addition property of equality to add the opposites of constants or variable expressions so that variable expressions are on one side of the equation and constants on the other.

3. Use the multiplication property of equality to multiply both sides by the reciprocal of the coefficient of the variable (that is, divide both sides by the coefficient) so that the new coefficient is 1.

4. Check your answer by substituting it into the _____ equation.

Example 2: Solving a Linear Equation

Solve $2(y - 7) = 4(y + 1) - 26$.

Solution

Example 3: Solving a Linear Equation

Solve $\dfrac{x - 4}{10} + \dfrac{7}{2} = \dfrac{x + 1}{5}$.

Solution

Example 5: Solutions of Equations

Determine whether each of the following equations is a conditional equation, an identity, or a contradiction.

a. $0.7 + 0.9x = 16$

Solution

b. $3x + 2(x + 5) = 5(x - 1)$

Solution

c. $5(x + 1) - 6x = -x + 5$

Solution

1.1 Exercises

1. Solve the following linear equation. Express your answer as an integer or a simplified fraction.

   $4y + 3 = 19$

2. Solve the following linear equation. Express your answer as an integer or a simplified fraction.

   $2 = -6t + 20$

3. Solve the following linear equation. Express your answer as an integer or a simplified fraction.

   $-5u - 24 = 7u$

4. Solve the following linear equation. Express your answer as an integer or a simplified fraction.

   $6v + 16 = -v + 16$

5. Solve the following linear equation. Express your answer as an integer or a simplified fraction.

   $-8x - 12 = 2\left(-4x - 6\right)$

6. Determine if the following linear equation has one solution, no solution, or if the solution is all real numbers. Express your solution as either a simplified fraction or decimal rounded to two places.

   $3 + \left(4z - 3\right) = 6\left(-6z + 5\right) + 4$

7. Determine if the following linear equation has one solution, no solution, or if the solution is all real numbers. Express your solution as either a simplified fraction or decimal rounded to two places.

   $-0.1t + 5.5 = 1.6t - 1.3$

8. Determine if the following linear equation has one solution, no solution, or if the solution is all real numbers. Express your solution as either a simplified fraction or decimal rounded to two places.

$$\frac{-2x}{3} = \frac{2x}{3} + \frac{4}{3}$$

9. Determine if the following linear equation has one solution, no solution, or if the solution is all real numbers. Express your solution as either a simplified fraction or decimal rounded to two places.

$$\frac{-5x}{2} + \frac{x+2}{3} = \frac{-3}{4}$$

10. Determine if the following linear equation has one solution, no solution, or if the solution is all real numbers. Express your solution as either a simplified fraction or decimal rounded to two places.

$$-\frac{3}{4}\left(z + \frac{1}{9}\right) = -\frac{1}{8}(z + 6)$$

1.2 Applications of Linear Equations in One Variable

EXAMPLE 1: Solving Linear Equations for One Variable

Solve the following equations for the specified variable. All of the equations are formulas that arise in application problems, and they are linear in the specified variable.

a. $P = 2l + 2w$. Solve for w.

b. $A = P\left(1 + \dfrac{r}{m}\right)^{mt}$. Solve for P.

c. $S = 2\pi r^2 + 2\pi rh$. Solve for h.

Solution

EXAMPLE 3: Calculating Average Interest

Julie invested \$1500 in a risky high-tech stock on January 1st. On July 1st, her stock is worth \$2100. She knows that her investment does not earn interest at a constant rate, but she wants to determine her average annual rate of return at this point in the year. What is the average annual rate of return she has earned so far?

Solution

1.2 Exercises

1. Solve the following formula for the indicated variable.

$C = 2\pi r$; solve for r.

2. Solve the following formula for the indicated variable.

$PV = nRT$; solve for T.

3. Solve the following formula for the indicated variable.

$A = \dfrac{1}{2}(B + b)\,h$; solve for b.

4. Solve the following formula for the indicated variable.

$C = \dfrac{5}{9}(F - 32)$; solve for F.

5. Solve the following formula for the indicated variable.

$A = 2lw + 2wh + 2hl$; solve for w.

6. Solve the following formula for the indicated variable.

$V = \dfrac{4}{3}\pi r^3$; solve for r.

7. Two trucks leave a warehouse at the same time. One travels due east at an average speed of 60 miles per hour, and the other travels due west at an average speed of 65 miles per hour. After how many hours will the two trucks be 625 miles apart?

8. Two cars leave a rest stop at the same time and proceed to travel down the highway in the same direction. One travels at an average rate of 46 miles per hour, and the other at an average rate of 54 miles per hour. How far apart are the two cars after 4.5 hours?

9. Two brothers, Mark and Walter, each inherit $17000. Mark invests his inheritance in a savings account with an annual return of 2.9%, while Walter invests his inheritance in a CD paying 4.4% annually. How much more money than Mark does Walter have after 1 year?

10. Troy buys a sound system priced at $8820, but pays $9702.00 with tax. What is the tax rate where Troy lives?

1.3 Linear Inequalities in One Variable

Cancellation Properties for Inequalities

In this table, A, B, and C represent algebraic expressions and D represents a nonzero constant. Each of the properties is stated for the inequality symbol $<$, but they are also true for the other three symbols (when substituted below).

Property	Description
If $A < B$, then $A + C < B + C$.	Adding the _____ to both sides of an inequality results in an equivalent inequality.
If $A < B$, and $D > 0$, then $A \cdot D < B \cdot D$.	If both sides of an inequality are multiplied by a _____ constant, the sense of the inequality is unchanged.
If $A < B$, and $D < 0$, then $A \cdot D > B \cdot D$.	If both sides are multiplied by a _____ constant, the sense of the inequality is reversed.

Example 2: Solving Linear Inequalities

Solve the following inequalities, using interval notation to describe the solution set.

a. $5 - 2(x - 3) \leq -(1 - x)$

b. $\dfrac{3(a - 2)}{2} < \dfrac{5a}{4}$

Solution

Example 3: Graphing Intervals of Real Numbers

a. $[-3, 6]$

b. $(-\infty, 5]$

c. $[2, 9)$

Solution

Example 6: Absolute Value Inequalities

Solve the following absolute value inequalities.

a. $|4 - 2x| > 6$

b. $2|3y - 2| + 3 \leq 11$

c. $|5 + 2s| \leq -3$

d. $|5 + 2s| \geq -3$

Solution

1.3 Exercises

1. Consider the following linear inequality.

$6y - 18 < -12 + 4y$

   a. Solve the inequality and express your answer in interval notation. Use decimal form for numerical values.

   b. Graph the solution set.

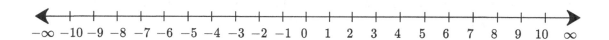

2. Consider the following linear inequality.

$$\frac{7y + 4.8}{8} > \frac{6y + 3.2}{7}$$

   a. Solve the inequality and express your answer in interval notation. Use decimal form for numerical values.

   b. Graph the solution set.

3. Consider the following linear inequality.

$-3.3z - 11.7 \geq 6.3\left(1 - z\right)$

   a. Solve the inequality and express your answer in interval notation. Use decimal form for numerical values.

   b. Graph the solution set.

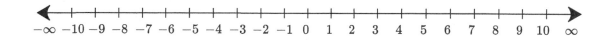

4. Consider the following compound inequality.

$$-36 < 4n - 8 \leq 4$$

   a. Solve the inequality and express your answer in interval notation. Use decimal form for numerical values.

   b. Graph the solution set.

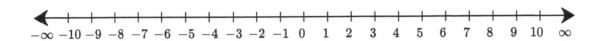

5. Consider the following compound inequality.

$$-1.75 \leq \frac{n+5}{-4} \leq 1$$

   a. Solve the inequality and express your answer in interval notation. Use decimal form for numerical values.

   b. Graph the solution set.

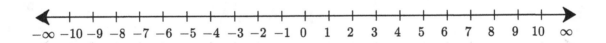

6. Consider the following inequality problem.

$$\frac{4t + 4}{4} > 10 \text{ and } -3\,(t - 3) > 27$$

   a. Solve the **first** inequality and express your answer in interval notation. Use decimal form for numerical values.

   b. Solve the **second** inequality and express your answer in interval notation. Use decimal form for numerical values.

   c. Using your answers from the previous steps, solve the **overall** inequality problem. Express your answer in interval notation. Use decimal form for numerical values.

7. Consider the following inequality problem.

$-5\,(n-2) \le 30$ or $7+n < 16$

    a. Solve the **first** inequality and express your answer in interval notation. Use decimal form for numerical values.

    b. Solve the **second** inequality and express your answer in interval notation. Use decimal form for numerical values.

    c. Using your answers from the previous steps, solve the **overall** inequality problem and express your answer in interval notation. Use decimal form for numerical values.

    d. Graph the **overall** solution set.

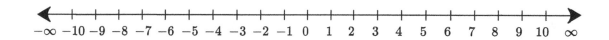

8. Jillian owns a orange grove, and needs to harvest at least 3225 oranges to cover the costs of running the grove. If each tree bears an average of 85 oranges and Jillian wants to donate 260 oranges to charity, how many orange trees need to be in the grove? Express your answer in interval notation.

9. Consider the following absolute value inequality.

$4\,|p+1| \le 4$

    a. Solve the inequality and express your answer in interval notation. Use decimal form for numerical values.

    b. Graph the solution set.

10. To qualify for Gold status at Awesome Airlines, one must fly at least 16200 and less than 36000 miles each year. If Greg takes a 900-mile round-trip flight to visit his parents, how many times does Greg need to visit his parents each year to attain Gold status? Express your answer in interval notation.

1.4 Quadratic Equations in One Variable

Quadratic Equations

A **quadratic equation in one variable**, say the variable x, is an equation that can be transformed into the form _____ , where a, b, and c are real numbers and $a \neq 0$. Such equations are also called _____ equations, as x appears to the second power. The name _____ comes from the Latin word *quadrus*, meaning "square".

Zero-Factor Property

Let A and B represent algebraic expressions. If the product of A and B is 0, then

_____ . That is,

$$AB = 0 \Rightarrow \underline{\hspace{4cm}}$$

Example 1: Solving Quadratic Equations by Factoring

Solve the following quadratic equations by factoring.

a. $x^2 + \dfrac{11x}{3} = \dfrac{4}{3}$

   Solution

b. $4x^2 + 28x = 0$

   Solution

Example 2: "Perfect Square" Quadratic Equations

Solve the following quadratic equations by taking square roots.

a. $(x - 6)^2 - 7 = 0$

   Solution

Method of Completing the Square

Step 1: Write the equation $ax^2 + bx + c = 0$ **in the form** _____ .

Step 2: Divide by _____ **, if** $a \neq 1$**, so that the coefficient of** x^2 **is 1:** $x^2 + \dfrac{b}{a}x = -\dfrac{c}{a}$.

Step 3: Divide the coefficient of *x* **by** _____ **, square the result, and** _____

_____ .

Step 4: The trinomial on the left side will now be _____ **. That is, it can be written as the square of an algebraic expression.**

Example 3: Completing the Square

Solve the following quadratic equations by completing the square.

a. $x^2 - 6x - 2 = 0$

   Solution

b. $25x^2 + 5x = 6$

Solution

The Quadratic Formula

The solutions of the general quadratic equation $ax^2 + bx + c = 0$, where $a \neq 0$ are

$$x = \underline{\hspace{6cm}} .$$

Example 4: The Quadratic Formula

Solve the following quadratic equations by using the quadratic formula.

a. $6y^2 - 5y = 1$

Solution

1.4 Exercises

1. Solve the following quadratic equation by factoring. If needed, write your answer as a fraction reduced to lowest terms.

$16x^2 - 121 = 0$

2. Solve the following quadratic equation by factoring. If needed, write your answer as a fraction reduced to lowest terms.

$(y - 9)^2 = 16$

3. Solve the following quadratic equation by factoring. If needed, write your answer as a fraction reduced to lowest terms.

$3y^2 + 9y = 0$

4. Solve the following quadratic equation by the square root method. If needed, write your answer as a fraction reduced to lowest terms.

$(y - 5)^2 = 9$

5. Solve the following quadratic equation by the square root method.

$-4(y + 6)^2 = -36$

6. Solve the following quadratic equation by completing the square.

$z^2 + 20z + 84 = 0$

7. Solve the following quadratic equation by completing the square.

$y^2 + 22y + 24 = -48$

8. Solve the following quadratic equation by using the quadratic formula. If needed, submit your answer as a fraction reduced to lowest terms.

$2y^2 - 5y = 59.5$

9. Solve the following quadratic equation by using the quadratic formula. If needed, submit your answer as a fraction reduced to lowest terms.

$3y^2 - 7y = 0$

10. Solve the following quadratic equation by using the quadratic formula. If needed, submit your answer as a fraction reduced to lowest terms.

$3x^2 + 8x - 3 = 6x$

1.5 Higher Degree Polynomial Equations

Quadratic-Like Equations

An equation is **quadratic-like**, or _____ , if it can be written in the form

$$aA^2 + bA + c = 0,$$

where a, b, and c are constants, $a \neq 0$, and A is an algebraic expression. Such equations can be solved by first _____ and then solving for the variable in the expression A. This method of solution is called _____ .

EXAMPLE 1: Quadratic-Like Equations

Solve the quadratic-like equations.

a. $\left(x^2 + 2x\right)^2 - 7\left(x^2 + 2x\right) - 8 = 0$

b. $y^{\frac{2}{3}} + 4y^{\frac{1}{3}} - 5 = 0$

Solution

1.5 Exercises

1. Solve the following quadratic-like equation.

$$\left(x^2 - 1\right)^2 - 15\left(x^2 - 1\right) + 56 = 0$$

2. Solve the following quadratic-like equation.

$$\left(x^2 - 6x\right)^2 + 2\left(x^2 - 6x\right) - 15 = 0$$

3. Solve the following quadratic-like equation.

$$2y^{\frac{2}{3}} + 16y^{\frac{1}{3}} + 14 = 0$$

4. Solve the following polynomial equation.

$$z^4 = 196$$

5. Solve the following polynomial equation.

$$z^4 + 3z^2 - 4 = 0$$

6. Solve the following polynomial equation.

$$y^3 + 27 = 0$$

7. Solve the following polynomial equation.

$$5y^3 + 28y^2 = 49y$$

8. Solve the following polynomial equation.

$$y^3 + 5y^2 - 16y - 80 = 0$$

Note: Some values may be repeated, as all input boxes will be used.

9. Solve the following equation by factoring.

$$5y^{\frac{8}{3}} + 28y^{\frac{5}{3}} - 12y^{\frac{2}{3}} = 0$$

10. Solve the following equation by factoring.

$$(z-3)^{\frac{-5}{2}} + 8(z-3)^{\frac{-3}{2}} = 0$$

1.6 Rational and Radical Equations

Proportion

A _____ is an equation stating that two ratios are equal.

Example 1: Proportions

Solve the following proportions.

b. $\dfrac{7}{x-2} = \dfrac{5}{x}$

Solution

To Solve an Equation Containing Rational Expressions

1. Find the _____ of the denominators.

2. Multiply both sides of the equation by this _____ and simplify.

3. Solve the resulting equation. (This equation will have only polynomials on both sides.)

4. Check each solution in the **original equation**. (Remember that no denominator can be _____ and any solution that gives a 0 _____ is to be discarded.)

Method for Solving Radical Equations

Step 1:

Begin by isolating the _____ on one side of the equation. If there is more than one radical expression, choose one to isolate on one side.

Step 2:

Raise both sides of the equation by the power necessary to "undo" the isolated Radical. That is, if the radical is an nth root, raise both sides to the nth power.

Step 3:

If any radical expressions remain, simplify the equation if possible and then repeat steps 1 and 2 until the result is a polynomial equation. When a polynomial equation has been obtained, solve the equation using polynomial methods.

Step 4:

Check your solutions in the _____ equation! Any extraneous solutions must be discarded.

Example 3: Radical Equations

Solve the following radical equations.

b. $\sqrt[3]{x^2 + 5x + 21} - 3 = 0$

Solution

1.6 Exercises

1. Consider the following equation:

$$\frac{-6}{x+8} = \frac{1}{-8}$$

    a. State any restriction(s) on the variable, if they exist.

    b. Solve the equation, if possible. If there is a solution, express your answer as either an integer or a simplified fraction.

2. Consider the following equation:

$$\frac{-x}{x+6} = \frac{-1}{6}$$

    a. State any restriction(s) on the variable, if they exist.

    b. Solve the equation, if possible. If there is a solution, express your answer as either an integer or a simplified fraction.

3. Consider the following equation:

$$\frac{-4}{x-4} = 1 - \frac{2}{x+8}$$

    a. State any restriction(s) on the variable, if they exist.

    b. Solve the equation, if possible. If there is a solution, express your answer as either an integer or a simplified fraction.

4. Consider the following equation:

$$\frac{x}{x-1} - \frac{3x}{x^2+x-2} = \frac{3}{x+2}$$

    a. State any restriction(s) on the variable, if they exist.

    b. Solve the equation, if possible. If there is a solution, express your answer as either an integer or a simplified fraction.

5. Consider the following equation:

$$\frac{x}{x+2} + \frac{3}{x+1} = 1$$

    a. State any restriction(s) on the variable, if they exist.

    b. Solve the equation, if possible. If there is a solution, express your answer as either an integer or a simplified fraction.

6. Solve the following radical equation. Separate multiple answers with a comma.

$$\sqrt{1-x} - x = 1$$

7. Solve the following radical equation. Separate multiple answers with a comma.

$$\sqrt[3]{5z^2 + 6z} = 3$$

8. Solve the following radical equation. Separate multiple answers with a comma.

$$\sqrt{2x+33} + 9 = x + 8$$

9. Solve the following equation. If needed, submit your answer as a fraction reduced to lowest terms. Separate multiple answers with a comma.

$$x^{\frac{3}{2}} - 125 = 0$$

10. Solve the following equation. If needed, submit your answer as a fraction reduced to lowest terms. Separate multiple answers with a comma.

$$z^{\frac{2}{3}} - \frac{1}{36} = 0$$

CHAPTER 2

Linear Equations in Two Variables

2.1 The Cartesian Coordinate System

Ordered Pairs

An _____ (a, b) consists of two real numbers a and b such that the _____ of a and b matters. That is, $(a, b) = (b, a)$ if and only if $a = b$. The number _____ is called the **first coordinate** and the number _____ is called the **second coordinate**.

The Cartesian Coordinate System

The **Cartesian coordinate system** (also called the **Cartesian plane**) consists of two perpendicular real number lines (each called an _____) intersecting at the 0 point of each line. The point of intersection is called the _____ of the system, and the four quarters defined by the two lines are called the _____ of the plane, numbered as indicated below in Figure 1. Because the Cartesian plane consists of two crossed real lines, it is often given the symbol $\mathbb{R} \times \mathbb{R}$, or $\mathbb{R}^2$. Each point P in the plane is identified by an _____. The first coordinate indicates the _____ displacement of the point from the origin, and the second coordinate indicates the _____ displacement. Figure 1 is an example of a Cartesian coordinate system and illustrates how several ordered pairs are **graphed**, or **plotted**.

CAUTION!

Unfortunately, mathematics uses parentheses to denote ordered pairs as well as open intervals, which sometimes leads to confusion. Context is the key to interpreting notation correctly. For instance, in the context of solving a one-variable inequality, the notation $(-2, 5)$ most likely refers to the _____ with endpoints at -2 and 5, while in the context of solving an equation in two variables, $(-2, 5)$ probably refers to a _____ in the Cartesian plane.

Distance Formula

The **distance** between two points (x_1, y_1) and (x_2, y_2) in the Cartesian plane is given by the following

EXAMPLE 3: Using the Distance Formula

Calculate the distance between the following pairs of points.

a. $(-4, -2)$ and $(-7, 2)$

b. $(5, 1)$ and $(-1, 3)$

Solution

Midpoint Formula

The **midpoint** between two points (x_1, y_1) and (x_2, y_2) in the Cartesian plane has the following coordinates:

$$\underline{\hspace{5cm}}$$

EXAMPLE 4: Using the Midpoint Formula

Calculate the midpoint of the line connecting each pair of points.

a. $(5, 1)$ and $(-1, 3)$

b. $(3, 0)$ and $(-6, 11)$

Solution

2.1 Exercises

1. Plot the following points in the Cartesian plane.

$$\{(7,2),(-3,6),(-4,-7),(6,-9)\}$$

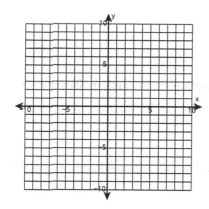

2. Identify the quadrant or axis that the following point lies on. If the point lies on an axis, specify which part (positive or negative) of which axis (x or y).

$(9,-5)$

3. For the following equation, determine the values of the missing entries. Reduce all fractions to lowest terms.

$9x - 6y = 16$

Note: Each column in the table represents an ordered pair. If multiple solutions exist, you only need to identify one.

x		0	2	
y	0			3

4. Consider the following pair of points.

$(7,-8)$ and $(-10,-4)$

  a. Determine the distance between the two points.

  b. Determine the midpoint of the line segment joining the pair of points.

5. Consider the following points.

$(3, -3)$, $(-4, 5)$ and $(-4, -3)$

    a. Determine whether or not the given points form a right triangle. If the triangle is **not** a right triangle, determine if it is isosceles or scalene.

    b. What is the area of the triangle? (Hint: Make use of the midpoint formula if the triangle is isosceles.)

6. For the following equation, determine the values of the missing entries. Reduce all fractions to lowest terms.

$y = x^2 - 12x + 36$

Note: Each column in the table represents an ordered pair. If multiple solutions exist, you only need to identify one.

x		5		8	3
y	0		1		

7. For the following equation, determine the values of the missing entries. Reduce all fractions to lowest terms.

$x^2 + y^2 = 81$

Note: Each column in the table represents an ordered pair. If multiple solutions exist, you only need to identify one.

x	0		$4\sqrt{5}$	$-\sqrt{77}$	
y		0			4

8. For the following equation, determine the values of the missing entries. Reduce all fractions to lowest terms.

$x = y^2$

Note: Each column in the table represents an ordered pair. If multiple solutions exist, you only need to identify one.

x	0		9	16	
y		2			$-\sqrt{7}$

9. Find the perimeter of the triangle whose vertices are the following specified points in the plane.

$(-10, -4), (-9, -7)$ and $(-8, 7)$

10. Your itinerary contains a grid map of New York, with each unit on the grid representing 0.125 miles. If the Rockefeller Center is located at $(4, -2)$ and Central Park is located at $(-11, 6)$, what is the direct distance (not walking distance, which would have to account for bridges and roadways) between the two landmarks in miles? Round your answer to two decimal places, if necessary.

2.2 Linear Equations in Two Variables

Linear Equations in Two Variables

A **linear equation in two variables,** say the variables x and y, is an equation that can be written in the form _____ , where a, b, and c are constants and a and b are not both zero. This form of such an equation is called the _____ **form.**

The x- and y-Intercepts

Given a graph in the Cartesian plane, any point where the graph intersects the x-axis is called an _____ and any point where the graph intersects the y-axis is called a

_____ .

All x-intercepts are of the form _____ and all y-intercepts are of the form _____ .

EXAMPLE 2: Finding Intercepts and Graphing Linear Equations

Find the x- and y-intercepts of the following equations, and then graph each equation.

a. $3x - 4y = 12$

Solution

2.2 Exercises

1. Determine if the following equation is linear. If the equation is linear, convert it to standard form: $ax + by = c$.

$9x + 4(x + 6y) = 4x + y$

2. Determine if the following equation is linear. If the equation is linear, convert it to standard form: $ax + by = c$.

$$-5x + 11(y + x) = 5$$

3. Determine if the following equation is linear. If the equation is linear, convert it to standard form: $ax + by = c$.

$$(7 + y)^2 - y^2 = 9x + 12$$

4. Determine if the following equation is linear.

$$5x^2 - (x - 2)^2 = y + 9$$

5. Determine if the following equation is linear.

$$9x + 12xy = 10y$$

6. Determine if the following equation is linear.

$$\frac{6}{x} + \frac{7}{y} = 11$$

7. Consider the following equation.

$$6x + y = 3$$

    a. Find the x- and y-intercepts, if possible.

b. Graph the equation by plotting the *x*- and *y*-intercepts. If an intercept does not exist, or is duplicated, use another point on the line to plot the graph.

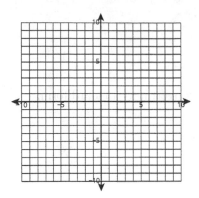

8. Consider the following equation.

$4y + 4 = x$

a. Find the *x*- and *y*-intercepts, if possible.

b. Graph the equation by plotting the *x*- and *y*-intercepts. If an intercept does not exist, or is duplicated, use another point on the line to plot the graph.

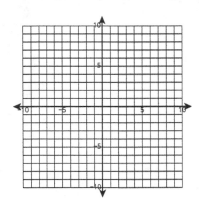

9. Consider the following equation.

$3y + 4x = 3y - 8$

a. Find the *x*- and *y*-intercepts, if possible.

b. Graph the equation by plotting the *x*- and *y*-intercepts. If an intercept does not exist, or is duplicated, use another point on the line to plot the graph.

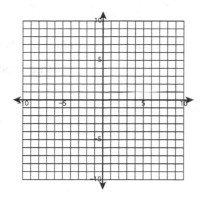

10. Consider the following equation.

$$-7y = -35$$

a. Find the *x*- and *y*-intercepts, if possible.

b. Graph the equation by plotting the *x*- and *y*-intercepts. If an intercept does not exist, or is duplicated, use another point on the line to plot the graph.

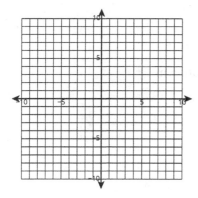

2.3 Forms of Linear Equations

The Slope of a Line

Let L stand for a given line in the Cartesian plane, and let (x_1, y_1) and (x_2, y_2) be the coordinates of any two distinct points on L. The **slope** of the line L is the ratio _____ which can be described in words as "change in y over change in x" or " _____ ."

CAUTION!

It doesn't matter how you assign the labels (x_1, y_1) and (x_2, y_2) to the two points you are using to calculate slope, but it *is* important that you are consistent as you apply the formula. You can not change the _____ in which you are subtracting as you determine the numerator and denominator in the slope formula.

Correct	Incorrect
$\dfrac{y_2 - y_1}{x_2 - x_1}$	$\dfrac{y_1 - y_2}{x_2 - x_1}$
$\dfrac{y_1 - y_2}{x_1 - x_2}$	$\dfrac{y_2 - y_1}{x_1 - x_2}$

Example 1: Calculating the Slope of a Line

Determine the slopes of the lines passing through the following pairs of points in $\mathbb{R}^2$.

   a. $(-4, -3)$ and $(2, -5)$

   b. $\left(\dfrac{3}{2}, 1 \right)$ and $\left(1, -\dfrac{4}{3} \right)$

   c. $(-2, 7)$ and $(1, 7)$

Solution

Slopes of Horizontal and Vertical Lines

Horizontal lines, which can be written in the form _____ , have a **slope of** _____ .

Vertical lines, which can be written in the form _____ , have an _____ **slope**.

Slope-Intercept Form of a Line

If the equation of a non-vertical line in x and y is solved for y, the result is an equation in **slope-intercept form**:

$$\underline{\hspace{4cm}} .$$

The constant m is the _____ of the line, and the _____ of the line is $(0, b)$. If the variable x does not appear in the equation, the slope is 0 and the equation is simply of the form $y = b$ and is a _____ .

Example 3: Slope-Intercept Form of a Line

Use the slope-intercept form of the line to graph the equation $4x - 3y = 6$.

Solution

Point-Slope Form of a Line

The **point-slope form** of the equation for the line passing through the point (x_1, y_1) with slope m is

$$\underline{\hspace{5cm}} .$$

Note that m, x_1, and y_1 are all constants, while x and y are variables. Note also that since the line, by definition, has slope m, _____ cannot be described in this form.

Example 6: Point-Slope Form of a Line

Find the equation, in slope-intercept form, of the line that passes through the two points $(-3, -2)$ and $(1, 6)$.

Solution

Standard Form: $ax + by = c$

Information Required: Typically, we arrive at the standard form when given a linear equation in another form.

Potential Uses: The standard form is most useful for easily calculating the _____ and

_____ .

Slope-Intercept Form: $y = mx + b$

Information Required: The slope m and the y-intercept $(0, b)$.

Potential Uses: The slope-intercept form makes it very easy to find the _____ and

_____ , and therefore to graph the line.

Point-Slope Form: $y - y_1 = m(x - x_1)$

Information Required: The slope m and a point on the line (x_1, y_1) or two points on the line (x_1, y_1) and (x_2, y_2).

Potential Uses: The point-slope form allows us to find the equation for a line when the

_____ is unknown.

2.3 Exercises

1. Find the slope of the line that passes through the following points. Simplify your answer.

$(-5, -2)$ and $(2, 0)$

2. Find the slope of the line determined by the following equation. Simplify your answer.

$-x + 6y = 1$

3. Find the slope of the line determined by the following equation. Simplify your answer.

$6y = 8$

4. Find the slope of the line determined by the following equation. Simplify your answer.

$-6x = 14$

5. Consider the following equation.

$4x + y = 6$

a. Express the given equation in slope-intercept form. Simplify your answer.

b. Given $x = 2$, determine the value for y.

c. Given $x = 3$, determine the value for y.

6. Consider the following equation.

$x + 3y - 2 = 0$

  a. Express the given equation in slope-intercept form. Simplify your answer.

  b. Given $x = -7$, determine the value for y.

  c. Given $x = -4$, determine the value for y.

7. Find the equation of the line in **slope-intercept form** that passes through the following point with the given slope. Simplify your answer.

Point $(9, -10)$; Slope $= -1$

8. Find the equation of the line in **slope-intercept form** that passes through the following point with the given slope. Simplify your answer.

Point $(-2, 11)$; Slope $= \dfrac{-1}{3}$

9. Find the equation of the line in **standard form** that passes through the following points. Simplify your answer.

$(10, 11)$ and $(10, 8)$

10. Sales at Glover's Golf Emporium have been increasing linearly. In their third business year, sales were $155,000. This year was their eighth business year, and sales were $310,000. If sales continue to increase at this rate, predict the sales in their tenth business year.

2.4 Parallel and Perpendicular Lines

Slopes of Parallel Lines

Two non-vertical lines with slopes m_1 and m_2 are **parallel** if and only if _____ . Also, two

vertical lines (with undefined slopes) are always _____ to each other.

EXAMPLE 2: Finding Equations of Parallel Lines

Find the equation, in slope-intercept form, for the line which is parallel to the line $3x + 5y = 23$ and which passes through the point $(-2, 1)$.

Solution

Slopes of Perpendicular Lines

Suppose m_1 and m_2 represent the slopes of two lines, neither of which is vertical. The two lines are

perpendicular if and only if _____ (equivalently, _____

and _____). If one of two perpendicular lines is vertical, the other is horizontal, and the slopes are, respectively, undefined and zero.

EXAMPLE 5: Finding Equations of Perpendicular Lines

Find the equation, in standard form, of the line that passes through the point $(-3, 13)$ and that is perpendicular to the line $y = -7$.

> **Note:**
>
> Remember that if you encounter a horizontal or vertical line, you cannot use the slope formulas to find a perpendicular line.

Solution

> **Note:**
>
> A pair of lines can not be *both* _____ and _____ .

2.4 Exercises

1. Consider the following equation.

$8x + 2y = 9$

  a. Express the given equation in slope-intercept form. Simplify your answer.

  b. Find the equation of the line which passes through the point $(-2, -10)$ and is **parallel** to the given line. Express your answer in slope-intercept form. Simplify your answer.

2. Consider the following equation.

$6y - 11 = -2(3 - 2x)$

  a. Express the given equation in slope-intercept form. Simplify your answer.

  b. Find the equation of the line which passes through the point $(-12, -4)$ and is **parallel** to the given line. Express your answer in slope-intercept form. Simplify your answer.

3. Consider the following equation.

$x + 6y = 2y - 3$

    a. Express the given equation in slope-intercept form. Simplify your answer.

    b. Find the equation of the line which passes through the point $(-9, -10)$ and is **perpendicular** to the given line. Express your answer in slope-intercept form. Simplify your answer.

4. Consider the following equation.

$3(y + x) - 8(x - y) = -7$

    a. Express the given equation in slope-intercept form. Simplify your answer.

    b. Find the equation of the line which passes through the point $(13, 5)$ and is **perpendicular** to the given line. Express your answer in slope-intercept form. Simplify your answer.

5. Consider the following equation.

$\dfrac{5x + 7}{3} - 4y = 2 - 3y$

    a. Express the given equation in slope-intercept form. Simplify your answer.

    b. Find the equation of the line which passes through the point $(1, -13)$ and is **perpendicular** to the given line. Express your answer in slope-intercept form. Simplify your answer.

6. Consider the following equations.

$-4y + 2x = 12$ and $3x - 2(x + 1) = 2y - 3x$

a. Express the **first** equation in slope-intercept form. Simplify your answer.

b. Express the **second** equation in slope-intercept form. Simplify your answer.

c. Determine if the two lines are **perpendicular**.

7. Given the following equations, determine if the lines are parallel, perpendicular, or neither.

$2 - (2y + 3x) = 4(x - y)$ and $2y + 4 = 7 + 7x$

8. Given the following equations, determine if the lines are parallel, perpendicular, or neither.

$\dfrac{6y + 7x}{5} = x + 2$ and $5x - 2y = 11x + 4$

9. Given the following equations, determine if the lines are parallel, perpendicular, or neither.

$\dfrac{5x - 3y}{4} = x + 1$ and $-y - x = 2x + 3$

2.5 Linear Regression

Scatter Plot

A _____ is a graphical display that is most commonly used to show two variables and how they might relate to one another. It is a graph on the coordinate plane that contains one point for each pair of data values.

Correlation

When data are plotted in a scatter plot and there appears to be an upward trend, that is, as one variable increases the other increases as well, we say there is a _____ between the variables. If the scatter plot trends downward, that is, as one variable increases the other decreases, we say there is a _____ between the variables.

Example 1: Identifying Correlations

Consider the relationship between the following variables and what kind of correlation might show up in a scatter plot of the data. Decide if the variables would likely have a positive correlation, negative correlation, or no linear correlation.

  a. The number of cigarettes smoked and the probability of lung cancer

  b. The number of minutes spent on social media sites by college students and their first semester grades

  c. The amount of credit card debt incurred by college freshmen and their IQ score

Solution

Pearson Correlation Coefficient

The **Pearson correlation coefficient**, rounded to the nearest thousandth, is a value between -1 and 1 that measures the strength of a linear correlation. For a sample, it is represented by the variable r.

Level of Confidence and Level of Significance

The _____ c is the probability that the assertions made about the data are correct.

The _____ α is the probability that the assertions made about the data are incorrect.

$$c + \alpha = 1$$

Statistically Significant

If $|r|$ is greater than the critical value listed in the table, then r is

_____ , which means it is unlikely to have occurred by chance.

Example 3: Identifying Statistical Significance

Use the table of the critical values of the Pearson correlation coefficient to determine if the correlation between BMI and age in Example 2 is statistically significant. Note that $r = 0.5763078191$. Use a 0.05 level of significance.

Solution

Regression Line

The _____ , also known as the _____ , is a particular line that most closely "fits" the data points on the scatter plot. The regression line can be represented by

$$\hat{y} = ax + b,$$

where a is the slope of the line and b is the y-intercept.

2.5 Exercises

1. Choose the graph that appears to have a positive linear correlation.

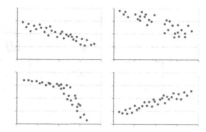

2. Which of the following scenarios is most likely to exhibit no correlation?

- Inches of rain and duration of thunderstorm

- Temperature and number of clothing layers worn

- Amount of homework completed and grade on quiz

- Running time of a movie and the munver of awards won

3. The following table shows the hours studied and corresponding test grade earned by students on a recent test. Calculate the correlation coefficient, r, and determine whether r is statistically significant at the 0.01 level of significance. Round your answer to the nearest thousandth.

Hours Studied and Test Grades								
Hours Studied	0.75	1	1	1.75	2.75	3	4.25	5.75
Test Grade	58	90	72	74	79	87	96	91

4. The following table gives the data for the grades on the midterm exam and the grades on the final exam. Determine the equation of the regression line, $\hat{y} = b_0 + b_1 x$. Round the slope and y-intercept to the nearest thousandth.

Grades on Midterm and Final Exams										
Grades on Midterm	63	62	72	99	80	80	76	89	85	70
Grades on Final	83	69	90	96	73	75	68	79	74	79

5. The following table gives the data for the average temperature and the snow accumulation in several small towns for a single month. Determine the equation of the regression line, $\hat{y} = b_0 + b_1 x$. Round the slope and y-intercept to the nearest thousandth. Then determine if the regression equation is appropriate for making predictions at the 0.05 level of significance.

Average Temperatures and Snow Accumulations										
Average Temperature (°F)	41	34	25	45	44	18	29	19	33	40
Snow Accumulation (in.)	7	13	23	10	15	25	21	11	13	9

6. The following table gives the data for the hours students spent on homework and their grades on the first test. The equation of the regression line for this data is $\hat{y} = 61.052 + 0.521x$. This equation is appropriate for making predictions at the 0.05 level of significance. If a student spent 35 hours on their homework, make a prediction for their grade on the first test. Round your prediction to the nearest whole number.

Hours Spent on Homework and Test Grades										
Hours Spent on Homework	34	12	43	37	12	29	49	52	8	30
Grade on Test	75	68	77	94	61	86	90	80	64	75

7. Use the table of critical values of the Pearson correlation coefficient to determine whether the correlation coefficient is statistically significant at the specified level of significance for the given sample size. Write the critical value to the nearest thousandth.

$r = 0.599, \alpha = 0.05, n = 12$

8. The following table gives the age and the number of hours a person watches TV per week. Estimate the correlation in words.

Age and Hours Watching TV per Week							
Age	8	14	29	40	42	55	61
Hours Watching TV	21	18	28	42	49	62	70

9. Use the linear regression model $\hat{y} = -17.8x + 806.72$ to predict the y-value for $x = 55$.

10. The following table gives the data for the grades on the midterm exam and the grades on the final exam. Determine the equation of the regression line, $\hat{y} = b_0 + b_1x$. Round the slope and y-intercept to the nearest thousandth.

Grades on Midterm and Final Exams									
Grades on Midterm	80	57	65	94	83	88	69	90	72	76
Grades on Final	90	68	83	99	78	78	72	89	73	63

CHAPTER 3

Functions and Their Graphs

3.1 Introduction to Functions

Function, Domain, and Range

Let D and R be two sets of real numbers. A _____ f is a rule that matches each number _____ in D with exactly one number _____ (or $f(x)$) in R. D is called the _____ of f, and R is called the _____ of f.

Note: In a function, there is only one y-value (or $f(x)$ value) for each value of x.

Example 2: Function Evaluation

For $f(x) = x^3 - 2x^2 + 3x + 100$, evaluate the following:

a. $f(2)$

Solution

b. $f(a)$

Solution

c. $f(a+1)$

Solution

d. $f(a) + 1$

Solution

Example 3: Function Evaluation

For $F(x) = x^2 - 8x$, evaluate the following:

a. $F(5) - F(2)$

Solution

b. $F(x + h)$

Solution

c. $F(x + h) - F(x)$

Solution

Example 4: Piecewise Function

A real estate broker charges a commission of 6% on sales valued up to $300,000. For sales valued at more than $300,000, the commission is $6000 plus 4% of the sale price.

a. Represent the commission earned as a function R.

Solution

b. Find $R(200,000)$

Solution

c. Find $R\,(500{,}000)$

Solution

Example 5: Domain

Find the domain of the following functions.

a. $f\,(x) = \dfrac{1}{x-2}$

Solution

b. $h\,(x) = \sqrt{x-2}$

Solution

Vertical Line Test

If any vertical line intersects a graph at more than _____ point, then the graph _____ represent a function.

3.1 Exercises

1. Given the following function:

$f\,(x) = 8x^2 - 3$

   a. Find $f\,(8)$.

   b. Find $f\,(-5)$.

   c. Find $f\,(x+h)$.

2. Given the following function:

$$f(x) = \frac{5x^2 - 1}{5x + 9}$$

   a. Find $f(3)$.

   b. Find $f(2) + f(3)$.

3. Given the following function:

$$f(x) = \begin{cases} 2x - 5 & \text{if } x < 8 \\ x^2 - 2 & \text{if } x \geq 8 \end{cases}$$

   a. Find $f(10)$.

   b. Find $f(9)$.

4. Find the domain of the following function:

$$f(x) = 9 + 3x^2$$

5. Find the domain of the following function:

$$f(x) = \begin{cases} 9x^2 - 7 & \text{if } x < 5 \\ 3x & \text{if } x \geq 5 \end{cases}$$

6. Find the domain of the following function:

$$f(x) = \frac{6}{\sqrt{7x + 10}}$$

7. Find the domain of the following function:

$$f(x) = \frac{3x + 10}{(x - 9)(x + 5)}$$

8. Determine if the graph shown below is a function:

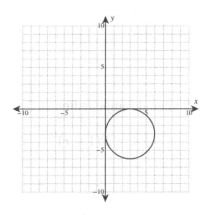

9. Find the domain of the following function:

$$f(x) = \frac{4x + 2}{x^2 - 4}$$

10. Find the domain of the following function:

$$f(x) = \sqrt{3x^2 + 2}$$

3.2 Functions and Models

Simple Interest

$$I = Prt$$

where $P =$ _____ , $r =$ _____ , and $t =$

_____ .

Compound Interest

A formula (model) for the balance with annual compounding at _____ r for t _____ is

$$A = P(1 + r)^t,$$

where r is in decimal form.

Example 1: Car Rental

Suppose that the total cost of renting a car consists of a fixed cost of $35 per day plus a variable cost of 15 cents per mile.

a. Write a cost function that represents the cost of driving x miles in one day.

b. Find the cost of driving 500 miles in one day.

Solution

Example 6: Physics

A basic law of physics states that, when an object is dropped in a near vacuum, the distance it falls varies directly with the square of the elapsed time. If an object dropped in a near vacuum falls 64 feet in 2 seconds, how far does it fall in 3 seconds? (Assume that the object does not hit the ground before 3 seconds have elapsed.)

Solution

Example 8: Phone Call Charges

Suppose that the cost of an overseas call is $9.00 for the first 3 minutes or less plus 95 cents for each additional minute. Write a function for the cost of a call of x minutes. (Assume that a fraction of a minute over 3 minutes is charged the corresponding fraction of 95 cents.)

Solution

3.2 Exercises

1. Find the break-even point(s) for the revenue and cost functions below. Separate multiple answers with a comma.

$$R(x) = 15x$$
$$C(x) = 11x + 8$$

2. Find the equilibrium point for the supply and demand functions below. Write your answer as an ordered pair.

$$S(x) = 9x + 8$$
$$D(x) = 148 - 5x$$

3. A record company has determined that the cost of producing CDs is $0.14 per CD plus $2780.40 per month in fixed costs. The record company sells each CD for $20.

   a. Find the revenue function.

   b. Find the cost function.

   c. Find the profit function.

   d. Find the break-even point.

4. The cost of producing 360 stereo speakers is $2540. Producing 640 stereo speakers would cost $4220.

    a. Find the average cost per stereo speaker of the additional 280 stereo speakers over 360.

    b. Assuming the total cost is a linear function, write an equation for the cost of producing x stereo speakers.

    c. What are the fixed costs?

5. Find the break-even point(s) for the revenue and cost functions below. Use exact form. Separate multiple answers with a comma.

$$R\left(x\right) = 24.2x - \frac{x^2}{5}$$
$$C\left(x\right) = 13x + 147$$

6. A bicycle tire inner tube producer can sell 29 inner tubes at a price of $1.45 per inner tube. If the price is $1.15, she can sell 41 inner tubes. The total cost of producing x inner tubes is $C\left(x\right) = 0.75x + 10.75$ dollars.

    a. Assuming the demand function is linear, find an equation for $D\left(x\right)$. Do not round your answer.

    b. Find the revenue function. Do not round your answer.

    c. Find the profit function. Do not round your answer.

7. For an international phone call, the phone company charges $10.75 for calls up to 5 minutes, plus $0.52 for each additional minute above 5. Find a function for the cost of a phone call where x is the length of a call in minutes.

8. It costs Thelma $14 to make a certain bracelet. She estimates that, if she charges x dollars per bracelet, she can sell $71 - 5x$ bracelets per week. Find a function for her weekly profit.

9. The perimeter of a rectangle is 254 inches. If the rectangle is x inches long, find a function for the area, $A(x)$.

10. The area of a rectangle is 220 m^2. If the length of the rectangle is x meters, write a function for the perimeter, $P(x)$.

3.3 Linear and Quadratic Functions

Linear Functions

A **linear function**, say f, of one variable, say the variable x, is any function that can be written in the form $f(x) =$ _____ , where m and b are real numbers. If $m \neq 0, f(x) = mx + b$ is also called a

_____ .

EXAMPLE 1: Graphing Linear Functions

Graph the following linear functions.

  a. $f(x) = 3x + 2$

  b. $g(x) = 3$

Solution

Quadratic Functions

A **quadratic function**, or _____ function, of one variable is any function that can be written in the form $f(x) = $ _____ , where a, b, and c are real numbers and $a \neq 0$.

Vertex and Axis of a Parabola

Figure 2 demonstrates two key characteristics of parabolas:

There is one point, known as the _____ , where the graph _____ .
Scanning the graph from left to right, it is the point where the graph stops going down and begins to go up (if the parabola opens upward) or stops going up and begins to go down (if the parabola opens downward).

Every parabola is _____ with respect to its **axis**, a line passing through the vertex dividing the parabola into two halves that are _____ of each other. This line is also called the **axis of symmetry**.

Vertex Form of a Quadratic Function

The graph of the function $g(x) = a(x - h)^2 + k$, where a, h, and k are real numbers and $a \neq 0$, is a parabola whose vertex is at _____ . The parabola is narrower than $f(x) = x^2$ if $|a|$ _____ 1, and is broader than $f(x) = x^2$ if 0 _____ $|a|$ _____ 1. The parabola opens upward if a is _____ and downward if a is _____ .

EXAMPLE 2: Graphing Quadratic Functions

Sketch the graph of the function $f(x) = -x^2 - 2x + 3$. Locate the vertex and the x-intercepts.

Solution

Vertex of a Quadratic Function

Given a quadratic function $f(x) = ax^2 + bx + c$, the graph of f is a parabola with a vertex given by:

$$\left(\underline{\hspace{3cm}} , \underline{\hspace{3cm}} \right) = \left(-\frac{b}{2a}, \frac{4ac - b^2}{4a} \right).$$

EXAMPLE 3: Using the Vertex Formula

Find the vertex of the following quadratic functions using the vertex formula.

a. $f(x) = x^2 - 4x + 8$

b. $g(x) = 3x^2 + 5x - 1$

Note:

If the x-coordinate of the vertex is simple, use substitution to find the y-coordinate. If the x-coordinate is complicated, use the explicit formula (the right hand form in the definition above).

Solution

3.3 Exercises

1. Consider the following function.

$$t(x) = (x - 4)^2 + 4$$

a. Find the vertex.

b. Find the x-intercepts, if any. Express the intercept(s) as ordered pair(s).

c. Find two points on the graph of the parabola other than the vertex and x-intercepts.

d. Graph the parabola.

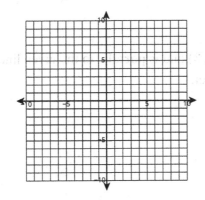

2. Consider the following function.

$$g(x) = (x - 3)(x + 5) + 16$$

a. Find the vertex.

b. Find the x-intercepts, if any. Express the intercept(s) as ordered pair(s).

c. Find two points on the graph of the parabola other than the vertex and x-intercepts.

d. Graph the parabola.

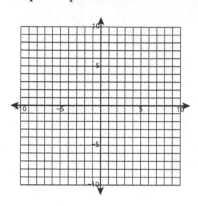

3. Consider the following equation.

$$6x + 3y = 11$$

   a. Express the given equation in slope-intercept form. Simplify your answer.

   b. Find the equation of the line which passes through the point $(-9, 3)$ and is **parallel** to the given line. Express your answer in slope-intercept form. Simplify your answer.

4. Consider the following equation.

$$5x + 10y = 9y - 4$$

   a. Express the given equation in slope-intercept form. Simplify your answer.

   b. Find the equation of the line which passes through the point $(-9, 4)$ and is **perpendicular** to the given line. Express your answer in slope-intercept form. Simplify your answer.

5. Find the linear function with the following properties.

$$f(0) = -5$$

Slope of $f = -4$

6. Consider the following function.

$$p(x) = 1 - \frac{1}{2}x$$

    a. Find the slope and the *y*-intercept. Express the intercept as an ordered pair. Simplify your answer.

    b. Plot two points on the line to graph the function.

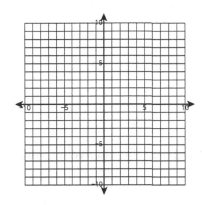

7. Consider the following function.

$$t(x) = (x - 4)^2 - 4$$

    a. Find the vertex.

    b. Find the *x*-intercepts, if any. Express the intercept(s) as ordered pair(s).

    c. Find two points on the graph of the parabola other than the vertex and *x*-intercepts.

d. Graph the parabola.

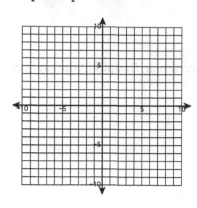

8. Consider the following function.

$$h\left(x\right) = 4\left(\frac{15}{7}x + \frac{1}{4}\right) - \frac{59}{7}x$$

a. Find the slope and the y-intercept. Express the intercept as an ordered pair. Simplify your answer.

b. Plot two points on the line to graph the function.

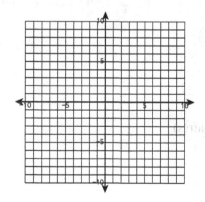

9. Consider the following function.

$$g\left(x\right) = x^2 + 6x + 5$$

a. Find the vertex.

b. Find the x-intercepts, if any. Express the intercept(s) as ordered pair(s).

c. Find two points on the graph of the parabola other than the vertex and x-intercepts.

d. Graph the parabola.

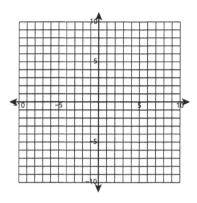

10. Consider the following function.

$$f(x) = -x^2 + 7x - 12$$

a. Find the vertex.

b. Find the x-intercepts, if any. Express the intercept(s) as ordered pair(s).

c. Find two points on the graph of the parabola other than the vertex and x-intercepts.

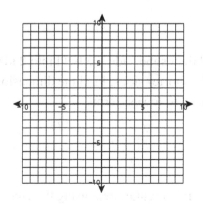

3.4 Applications of Quadratic Functions

EXAMPLE 1: Fencing a Garden

A farmer plans to use 100 feet of spare fencing material to form a rectangular garden plot against the side of a long barn, using the barn as one side of the plot. How should he split up the fencing among the other three sides in order to maximize the area of the garden plot?

Solution

3.4 Exercises

1. Among all rectangles that have a perimeter of 194, find the dimensions of the one whose area is largest. Write your answers as fractions reduced to lowest terms.

2. Among all pairs of numbers (x, y) such that $8x + 2y = 18$, find the pair for which the sum of squares, $x^2 + y^2$, is minimum. Write your answers as fractions reduced to lowest terms.

3. A rancher has 800 feet of fencing to put around a rectangular field and then subdivide the field into 2 identical smaller rectangular plots by placing a fence parallel to one of the field's shorter sides. Find the dimensions that maximize the enclosed area. Write your answers as fractions reduced to lowest terms.

4. The back of Alisha's property is a creek. Alisha would like to enclose a rectangular area, using the creek as one side and fencing for the other three sides, to create a pasture. If there is 600 feet of fencing available, what is the maximum possible area of the pasture?

5. The total revenue for Jane's Vacation Rentals is given as the function $R(x) = 300x - 0.5x^2$, where x is the number of apartments filled. What number of apartments filled produces the maximum revenue?

6. The total cost of producing a type of truck is given by $C(x) = 23000 - 90x + 0.1x^2$, where x is the number of trucks produced. How many trucks should be produced to incur minimum cost?

7. The revenue function for a lawnmower shop is given by $R(x) = x \cdot p(x)$ dollars where x is the number of units sold and $p(x) = 400 - 0.5x$ is the unit price. Find the maximum revenue.

8. A projectile is launched upward with a velocity of 192 feet per second from the top of a 90-foot stage. What is the maximum height attained by the projectile?

9. Determine whether the following function has a maximum, a minimum, or neither. If it has either a maximum or a minimum, find what that value is and where it occurs. Reduce all fractions to lowest terms.

$f(x) = -x^2 - 6x - 4$

10. Determine whether the following function has a maximum, a minimum, or neither. If it has either a maximum or a minimum, find what that value is and where it occurs. Reduce all fractions to lowest terms.

$f(x) = 7x + 5$

3.5 Other Common Functions

EXAMPLE 1: Functions of the Form ax^n

Sketch the graphs of the following functions.

a. $f(x) = \dfrac{x^4}{5}$

b. $g(x) = -x^3$

Solution

EXAMPLE 2: Functions of the Form $\dfrac{a}{x^n}$

Sketch the graph of the function $f(x) = -\dfrac{1}{4x}$.

Solution

EXAMPLE 3: The Absolute Value Function

Sketch the graph of the function $f(x) = -2\,|x|$.

Solution

The Greatest Integer Function

The greatest integer function, $f(x) = [\![x]\!]$, is a function commonly encountered in computer science applications. It is defined as follows: the **greatest integer of x** is the _____ integer less than or equal to x. For instance, $[\![4.3]\!] = 4$ and $[\![-2.9]\!] = -3$ (note that -3 is the largest integer to the left of -2.9 on the real number line).

Piecewise-Defined Function

A **piecewise-defined** function is a function defined in terms of _____ formulas, each valid for its own unique portion of the real number line. In evaluating a piecewise-defined function f at a certain value for x, it is important to correctly identify which formula is valid for that particular value.

EXAMPLE 4: Piecewise-Defined Function

Sketch the graph of the function $f(x) = \begin{cases} -2x - 2 & \text{if } x \leq -1 \\ x^2 & \text{if } x > -1 \end{cases}$.

Note:

Always pay close attention to the boundary points of each interval. Remember that only one rule applies at each point.

Solution

3.5 Exercises

1. Consider the following function.

$$h\left(x\right) = \frac{3}{2}x^3$$

    a. Identify the general shape of the graph of this function.

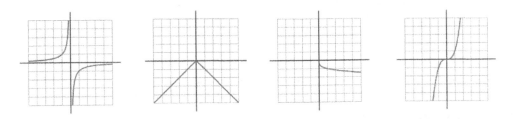

    b. Find two points on the graph of this function, other than the origin. Express each coordinate as an integer or simplified fraction, or round to four decimal places as necessary.

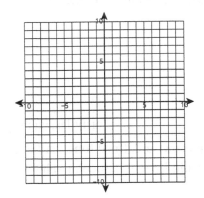

2. Consider the following function.

$$u\left(x\right) = -\frac{9}{7}x^4$$

    a. Identify the general shape of the graph of this function.

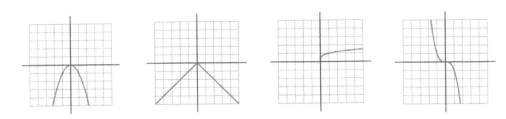

b. Find two points on the graph of this function, other than the origin. Express each coordinate as an integer or simplified fraction, or round to four decimal places as necessary.

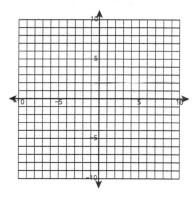

3. Consider the following function.

$$q(x) = -\frac{1}{2x}$$

a. Identify the general shape of the graph of this function.

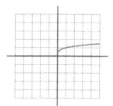 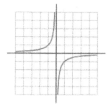 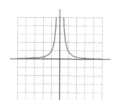 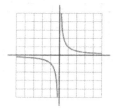

b. Find two points on the graph of this function, other than the origin. Express each coordinate as an integer or simplified fraction, or round to four decimal places as necessary.

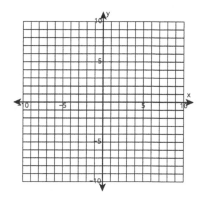

4. Consider the following function.

$$v(x) = -\frac{6}{5x^4}$$

a. Identify the general shape of the graph of this function.

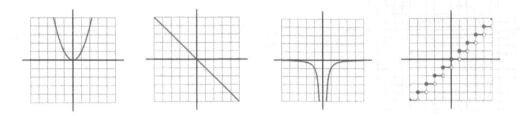

b. Find two points on the graph of this function, other than the origin. Express each coordinate as an integer or simplified fraction, or round to four decimal places as necessary.

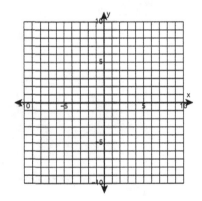

5. Consider the following function.

$$t(x) = \frac{\sqrt[3]{x}}{5}$$

a. Identify the general shape of the graph of this function.

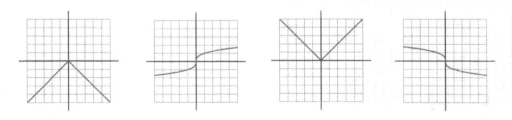

b. Find two points on the graph of this function, other than the origin. Express each coordinate as an integer or simplified fraction, or round to two decimal places as necessary.

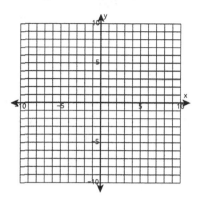

6. Consider the following function.

$$u(x) = -\frac{4\sqrt[4]{x}}{3}$$

a. Identify the general shape of the graph of this function.

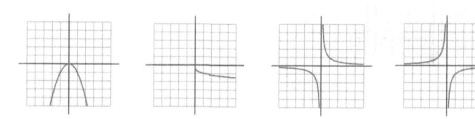

b. Find two points on the graph of this function, other than the origin. Express each coordinate as an integer or simplified fraction, or round to two decimal places as necessary.

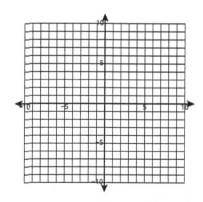

7. Consider the following function.

$$f(x) = -\frac{5}{9}|x|$$

    a. Identify the general shape of the graph of this function.

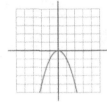 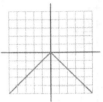 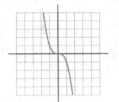

    b. Find two points on the graph of this function, other than the origin. Express each coordinate as an integer or simplified fraction.

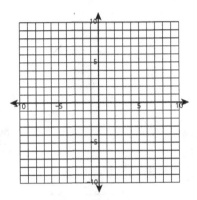

8. Consider the following function.

$$p(x) = 3\left[\frac{-x}{3}\right]$$

    a. Identify the general shape of the graph of this function.

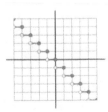 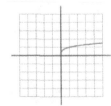 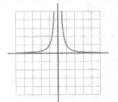 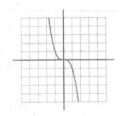

b. Find the length of the individual line segments of this function. Then, find the positive vertical separation between each line segment. Simplify your answer.

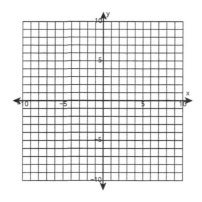

9. Graph the following piecewise-defined function.

$$w(x) = \begin{cases} -2x - 4 & \text{if } x \leq -2 \\ \dfrac{7}{9}x^2 & \text{if } x > -2 \end{cases}$$

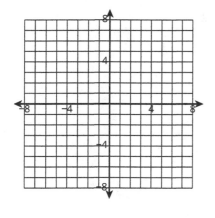

10. Graph the following piecewise-defined function.

$$t\left(x\right) = \begin{cases} -\dfrac{5}{8}\left|x\right| & \text{if } x < 1 \\ 2\left[\!\left[x\right]\!\right] & \text{if } x \geq 1 \end{cases}$$

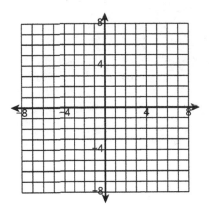

3.6 Transformations of Functions

Theorem: Horizontal Shifting/Translation

Let $f(x)$ be a function, and let h be a fixed real number. If we replace x with $x - h$, we obtain a new function $g(x) = f(x - h)$. The graph of g has the same _____ as the graph of f, but shifted to the _____ by h units if $h > 0$ and shifted to the _____ by h units if $h < 0$.

EXAMPLE 1: Horizontal Shifting/Translation

Sketch the graphs of the following functions.

a. $f(x) = (x + 2)^3$

b. $g(x) = |x - 4|$

Solution

CAUTION!

The minus sign in the expression $x - h$ is critical. When you see an expression in the form $x + h$ you must think of it as _____.

Consider a specific example: replacing x with $x - 5$ shifts the graph 5 units to the *right*, since 5 is positive. Replacing x with $x + 5$ shifts the graph 5 units to the _____, since we have actually replaced x with $x - (-5)$.

Vertical Shifting/Translation

Let $f(x)$ be a function whose graph is known, and let k be a fixed real number. The graph of the function $g(x) = f(x) + k$ is the same shape as the graph of f, but shifted _____ if $k > 0$ and _____ if $k < 0$.

EXAMPLE 2: Vertical Shifting/Translation

Sketch the graphs of the following functions.

a. $f(x) = \dfrac{1}{x} + 3$

b. $g(x) = \sqrt[3]{x} - 2$

Solution

EXAMPLE 3: Horizontal and Vertical Shifting

Sketch the graph of the function $f(x) = \sqrt{x+4} + 1$.

> **Note:**
>
> In this case, it doesn't matter which shift we apply first. However, when functions get more complicated, it is usually best to apply _____ shifts before _____ shifts.

Solution

Reflecting With Respect to the Axes

Given a function $f(x)$:

1. The graph of the function $g(x) = -f(x)$ is the reflection of the graph of f with respect to the _____.

2. The graph of the function $g(x) = f(-x)$ is the reflection of the graph of f with respect to the _____.

In other words, a function is reflected with respect to the x-axis by multiplying the entire function by _____, and reflected with respect to the y-axis by replacing x with _____.

EXAMPLE 4: Reflecting With Respect to the Axes

Sketch the graphs of the following functions.

a. $f(x) = -x^2$

b. $g(x) = \sqrt{-x}$

Solution

Vertical Stretching and Compressing

Let $f(x)$ be a function and let a be a positive real number.

1. The graph of the function $g(x) = a f(x)$ is _____ compared to the graph of f if $a > 1$.

2. The graph of the function $g(x) = a f(x)$ is _____ compared to the graph of f if $0 < a < 1$.

EXAMPLE 5: Vertical Stretching and Compressing

Sketch the graphs of the following functions.

a. $f(x) = \dfrac{\sqrt{x}}{10}$

b. $g(x) = 5|x|$

Note:

When graphing stretched or compressed functions, it may help to plot a few points of the new function.

Solution

Order of Transformations

If a function g has been obtained from a simpler function f through a number of transformations, g can be understood by looking for transformations in this order:

1. _____

2. _____

3. _____

4. _____

EXAMPLE 6: Order of Transformations

Sketch the graph of the function $f(x) = \dfrac{1}{2-x}$.

Solution

y-axis Symmetry

The graph of a function f has **y-axis symmetry**, or is **symmetric with respect to the y-axis**, if $f(-x) = $ _____ for all x in the domain of f. Such functions are called _____ functions.

Origin Symmetry

The graph of a function f has **origin symmetry**, or is **symmetric with respect to the origin**, if $f(-x) = $ _____ for all x in the domain of f. Such functions are called _____ functions.

Symmetry of Equations

We say that an equation in x and y is **symmetric with respect to**:

1. the _____ if replacing x with $-x$ results in an equivalent equation

2. the _____ if replacing y with $-y$ results in an equivalent equation

3. the _____ if replacing x with $-x$ and y with $-y$ results in an equivalent equation

EXAMPLE 7: Symmetry of Equations

Sketch the graphs of the following relations, making use of symmetry.

a. $f(x) = \dfrac{1}{x^2}$

b. $g(x) = x^3 - x$

c. $x = y^2$

> **Note:**
>
> If you don't know where to begin when sketching a graph, plotting points often helps you understand the basic shape.

Solution

Increasing, Decreasing and Constant

We say that a function f is:

1. _____ **on an interval** if, for any x_1 and x_2 in the interval with $x_1 < x_2$, it is the case that $f(x_1)$ _____ $f(x_2)$.

2. _____ **on an interval** if, for any x_1 and x_2 in the interval with $x_1 < x_2$, it is the case that $f(x_1)$ _____ $f(x_2)$.

3. _____ **on an interval** if, for any x_1 and x_2 in the interval with $x_1 < x_2$, it is the case that $f(x_1)$ _____ $f(x_2)$.

EXAMPLE 8

Determine the open intervals of monotonicity of the function $f(x) = (x - 2)^2 - 1$.

Solution

3.6 Exercises

1. Consider the following function.

$$s(x) = (x - 3)^3$$

   a. Determine the more basic function that has been shifted, reflected, stretched, or compressed.

b. Graph the given function by indicating how the more basic function has been shifted, reflected, stretched, or compressed.

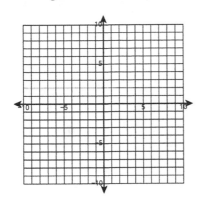

c. Determine the domain and range of the given function. Express your answer in interval notation.

2. Consider the following function.

$$v(x) = \frac{x^2}{3} - 2$$

a. Determine the more basic function that has been shifted, reflected, stretched, or compressed.

b. Graph the given function by indicating how the more basic function has been shifted, reflected, stretched, or compressed.

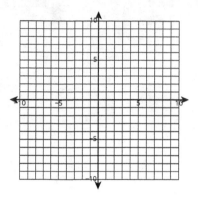

c. Determine the domain and range of the given function. Express your answer in interval notation.

3. Consider the following function.

$$v(x) = -\frac{1}{x-5}$$

a. Determine the more basic function that has been shifted, reflected, stretched, or compressed.

b. Graph the given function by indicating how the more basic function has been shifted, reflected, stretched, or compressed.

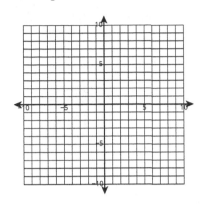

c. Determine the domain and range of the given function. Express your answer in interval notation.

4. Consider the following function.

$$q(x) = \sqrt{x+4}+1$$

a. Determine the more basic function that has been shifted, reflected, stretched, or compressed.

b. Graph the given function by indicating how the more basic function has been shifted, reflected, stretched, or compressed.

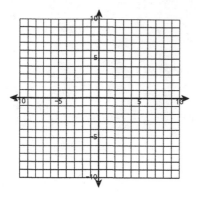

c. Determine the domain and range of the given function. Express your answer in interval notation.

5. Consider the following function.

$$s(x) = -2\sqrt{x+5} + 1$$

a. Determine the more basic function that has been shifted, reflected, stretched, or compressed.

b. Graph the given function by indicating how the more basic function has been shifted, reflected, stretched, or compressed.

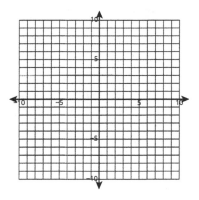

c. Determine the domain and range of the given function. Express your answer in interval notation.

6. Consider the following function.

$$n\left(x\right) = \frac{\sqrt{-x}}{2} + 4$$

a. Determine the more basic function that has been shifted, reflected, stretched, or compressed.

b. Graph the given function by indicating how the more basic function has been shifted, reflected, stretched, or compressed.

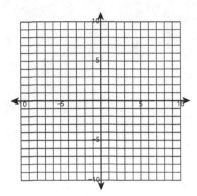

c. Determine the domain and range of the given function. Express your answer in interval notation.

7. Determine if the following function is even, odd, or neither.

$$u\left(x\right) = -\frac{x^2}{3} - 2$$

8. Determine if the following equation has x-axis symmetry, y-axis symmetry, origin symmetry, or none of these.

$$x + y^2 = 3$$

9. Determine if the following equation has x-axis symmetry, y-axis symmetry, origin symmetry, or none of these.

$$y = |2x|$$

10. Find the interval(s) where the following function is increasing, decreasing, or constant. Express your answer in interval notation.

3.7 Polynomial Functions

Zeros of a Polynomial

The number k (k may be a complex number) is a _____ of the polynomial function $p(x)$ if $p(k) = 0$. This is also expressed by saying that k is a _____ of the polynomial or a _____ of the equation $p(x) = 0$.

Polynomial Equations

A **polynomial equation in one variable**, say the variable x, is an equation that can be written in the form $a_n x^n + a_{n-1} x^{n-1} + \ldots + a_1 x + a_0 = 0$, where n is a nonnegative integer, $a_n, a_{n-1}, \ldots, a_1, a_0$ are _____ . Assuming $a_n \neq 0$, we say such an equation is of _____ n and call a_n the _____ .

EXAMPLE 1: Solutions of Polynomial Equations

Verify that the given values of x solve the corresponding polynomial equations.

a. $6x^2 - x^3 = 12 + 5x; x = 4$

Solution

Behavior of Polynomials as $x \to \pm\infty$

Given a polynomial function $p(x)$ with degree n, the behavior of $p(x)$ as $x \to \pm\infty$ can be determined from the leading term $a_n x^n$ using the table below.

	***n* is even**	***n* is odd**
a_n is positive	as $x \to -\infty$, $p(x) \to +\infty$ as $x \to +\infty$, $p(x) \to +\infty$ The graph _____ to the left and _____ to the right.	as $x \to -\infty$, $p(x) \to -\infty$ as $x \to +\infty$, $p(x) \to +\infty$ The graph _____ to the left and _____ to the right.
a_n is negative	as $x \to -\infty$, $p(x) \to -\infty$ as $x \to +\infty$, $p(x) \to -\infty$ The graph _____ to the left and _____ to the right.	as $x \to -\infty$, $p(x) \to +\infty$ as $x \to +\infty$, $p(x) \to -\infty$ The graph _____ to the left and _____ to the right.

EXAMPLE 2: Graphing Polynomial Functions

Sketch the graphs of the following polynomial functions, paying particular attention to the x-intercept(s), the y-intercept, and the behavior as $x \to \pm\infty$

a. $f(x) = -x\,(2x + 1)\,(x - 2)$

c. $h(x) = x^4 - 1$

Solution

Polynomial Inequalities

A **polynomial** _____ is any inequality that can be written in the form:

$$p(x) < 0, p(x) \leq 0, p(x) > 0, \text{ or } p(x) \geq 0,$$

where $p(x)$ is a polynomial function.

Solving Polynomial Inequalities: Sign-Test Method

To solve a polynomial inequality $p(x) < 0$, $p(x) \leq 0$, $p(x) > 0$, or $p(x) \geq 0$:

Step 1: Find the real _____ of $p(x)$.

Equivalently, find the real solutions of $p(x) = 0$.

Step 2: Place the zeros on a _____ , splitting it into intervals.

Step 3: Within each interval, select a _____ and evaluate p at that number.

If the result is _____ , then $p(x) > 0$ for all x in the interval. If the result is _____ , then $p(x) < 0$ for all x in the interval.

Step 4: Write the solution set, consisting of all of the intervals that satisfy the given inequality.

If the inequality is _____ (uses $\leq$ or $\geq$), then the zeros _____ included in the solution set as well.

EXAMPLE 4: Solving Polynomial Inequalities: Sign-Test Method

Solve the polynomial inequality $(x - 2)(x + 5)(x - 4) \leq 0$.

Solution

3.7 Exercises

1. Solve the following polynomial equation by factoring or using the quadratic formula. Identify all solutions.

$x^2 - 10x + 37 = 0$

2. Solve the following polynomial equation by factoring or using the quadratic formula. Identify all solutions.

$x^4 - 3x^2 + 2 = 0$

3. Graph the following polynomial function and identify the x- and y-intercepts. Express all points as ordered pairs.

$g(x) = x^3 + 5x^2 + 6x$

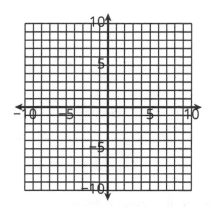

4. Graph the following polynomial function and identify the x- and y-intercepts. Express all points as ordered pairs.

$g(x) = (x - 5)(x + 1)(4 - x)$

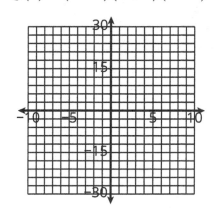

5. Graph the following polynomial function and identify the x- and y-intercepts. Express all points as ordered pairs.

$$r(x) = (x+6)(x+1)(x-2)(1-x)$$

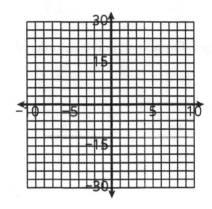

6. Find the solution of the following polynomial inequality. Express your answer in interval notation.

$$(x+2)(x-2)(5+x) > 0$$

7. Find the solution of the following polynomial inequality. Express your answer in interval notation.

$$x(x+5)(x-2) \leq 0$$

8. Consider the following polynomial.

$$g(x) = 6x^3 + 5x^2 - 2x + 10$$

  a. Find the degree and leading coefficient of $g(x)$.

  b. Describe the behavior of the graph of $g(x)$ as $x \to \pm\infty$.

9. Find the solution of the following polynomial inequality. Express your answer in interval notation.

$$x^2 \geq -x + 20$$

10. Find the solution of the following polynomial inequality. Express your answer in interval notation.

$x\,(x+4)\,(x-2) > 0$

3.8 Rational Functions

Rational Functions

A **rational function** is a function that can be written in the form

$$f(x) = \frac{\rule{3em}{0.4pt}}{\rule{6em}{0.4pt}} \; ,$$

where $p(x)$ and $q(x)$ are polynomial functions and $q(x) \neq 0$. Even though q is not allowed to be identically zero, there will often be values of x for which $q(x)$ is zero, and at these values the function is _____ . Consequently, the **domain of f** consists of all real numbers except those for which _____ .

Vertical Asymptotes

The vertical line $x = c$ is a _____ of a function f if $f(x)$ increases in magnitude without bound as x approaches c. Examples of vertical asymptotes appear in Figure 2. The graph of a rational function cannot _____ a vertical asymptote.

Horizontal Asymptotes

The horizontal line $y = c$ is a _____ of a function f if $f(x)$ approaches the value c as $x \to -\infty$ or as $x \to \infty$. Examples of horizontal asymptotes appear in Figure 3. The graph of a rational function may _____ a horizontal asymptote near _____ , but will eventually _____ the asymptote from one side only as x increases in magnitude.

Oblique Asymptotes

A non-vertical, non-horizontal line may also be an asymptote of a function f. Examples of _____ (or **slant**) **asymptotes** appear in Figure 4. Again, the graph of a rational function may _____ an oblique asymptote near _____ , but will eventually _____ the asymptote from one side only as $x \to \infty$ or $x \to -\infty$.

Asymptote Notation

The notation $x \to c^-$ is used when describing the behavior of a graph as x approaches the value c from the _____ (the negative side). The notation $x \to c^+$ is used when describing behavior as x approaches c from the _____ (the positive side). The notation $x \to c$ is used when describing behavior that is the same on _____ of c.

Equations for Vertical Asymptotes

If the rational function $f(x) = \dfrac{p(x)}{q(x)}$ has been written in reduced form (so that p and q have no common factors), the vertical line $x = c$ is a _____ of f if and only if c is a zero of the polynomial q. In other words, f has vertical asymptotes at the _____ of q.

EXAMPLE 1: Vertical Asymptotes

Find the domains and the equations for the vertical asymptotes of the following functions.

a. $f(x) = \dfrac{32}{x+2}$

b. $g(x) = \dfrac{x^2 + 1}{x^2 + 2x - 15}$

c. $h(x) = \dfrac{x^2 - x}{x - 1}$

Note:

We must always find the domain before canceling common factors. Even if a zero is removed from the denominator when finding the reduced form, that value is not part of the domain.

Solution

Equations for Horizontal and Oblique Asymptotes

Let $f(x) = \dfrac{p(x)}{q(x)}$ be a rational function, where p is an nth degree polynomial with leading coefficient a_n and q is an mth degree polynomial with leading coefficient b_m, and $p(x)$ and $q(x)$ have no common factors other than constants. Then the asymptotes of f are found as follows:

1. If $n < m$, the horizontal line _____ (the x-axis) is the **horizontal asymptote** for f.

2. If $n = m$, the horizontal line _____ is the **horizontal asymptote** for f.

3. If $n = m + 1$, the line $y = g(x)$ is an _____ for f, where g is the quotient polynomial obtained by _____ p by q.

4. If $n > m + 1$, there is _____ straight line _____ _____ for f.

EXAMPLE 2: Horizontal and Oblique Asymptotes

Find the equation for the horizontal or oblique asymptote of the following functions.

a. $f(x) = \dfrac{x^2 + 1}{x^2 + 2x - 15}$

b. $g(x) = \dfrac{x^3 + x^2 + 2x + 2}{x^2 + 9}$

c. $h(x) = \dfrac{3x^4 + 10x - 7}{x^6 + x^5 - x^2 - 1}$

d. $j(x) = \dfrac{2x^4 - 3x^2 + 8}{x^2 - 25}$

Note:

Always begin by comparing the degrees of the numerator and the denominator.

Solution

Graphing Rational Functions

Given a rational function f,

Step 1: _____ in order to determine the domain of f. Any

points excluded from the domain may appear as _____ in the graph or as

_____ .

Step 2: Factor the numerator as well and _____ any common factors.

Step 3: Examine the remaining factors in the denominator to determine the equations for any

_____ .

Step 4: Compare the degrees of the numerator and denominator to determine if there is a

_____ or _____ asymptote. If so, find its equation.

Step 5: Determine the _____ , if 0 is in the domain of f.

Step 6: Determine the _____ , if there are any, by setting the numerator of the
reduced fraction equal to 0.

Step 7: Plot enough points to determine the _____ of f between x-intercepts and
between vertical asymptotes.

EXAMPLE 3: Graphing Rational Functions

Sketch the graphs of the following rational functions.

a. $f(x) = \dfrac{x^2 - x}{x - 1}$

b. $g(x) = \dfrac{x^2 + 1}{x^2 + 2x - 15}$

c. $h(x) = \dfrac{x^3 + x^2 + 2x + 2}{x^2 + 9}$

Solution

3.8 Exercises

1. Find equations for the horizontal asymptotes, if any, for the following rational function.

$$f(x) = \frac{-5}{x - 4}$$

2. Find equations for the vertical asymptotes, if any, for the following rational function.

$$f(x) = \frac{-3x - 1}{x + 4}$$

3. Find equations for the horizontal asymptotes, if any, for the following rational function.

$$f(x) = \frac{-x + 8}{x + 4}$$

4. Find equations for the vertical asymptotes, if any, for the following rational function.

$$f(x) = \frac{3x^2 + 11x - 70}{x + 7}$$

5. Find equations for the vertical asymptotes, if any, for the following rational function.

$$f(x) = \frac{-4x^2 - 6x + 10}{-2x - 7}$$

6. Find equations for the oblique asymptotes, if any, for the following rational function.

$$f(x) = \frac{-4x^2 + 7x + 15}{4x + 9}$$

7. Graph the following rational function.

$$f(x) = \frac{-2}{x+1}$$

    a. Plot the vertical asymptotes, if any, on the graph.

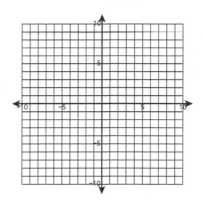

    b. Plot the horizontal asymptotes, if any, on the graph.

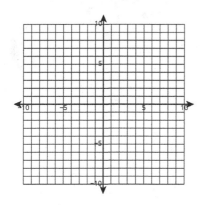

    c. Sketch the graph of the function.

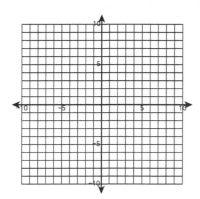

8. Graph the following rational function.

$$f(x) = \frac{-2}{x^2 - 64}$$

a. Plot the vertical asymptotes, if any, on the graph.

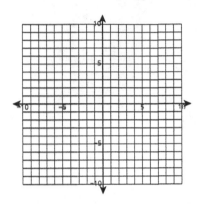

b. Plot the horizontal asymptotes, if any, on the graph.

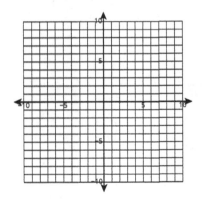

c. Sketch the graph of the function.

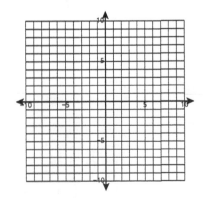

9. Graph the following rational function.

$$f(x) = \frac{-x - 3}{x^2 - 49}$$

    a. Plot the vertical asymptotes, if any, on the graph.

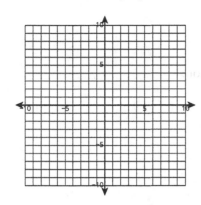

    b. Plot the horizontal asymptotes, if any, on the graph.

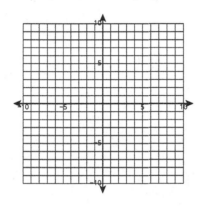

    c. Sketch the graph of the function.

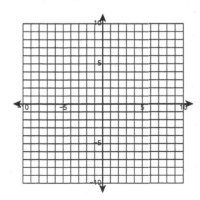

10. Consider the following rational function.

$$f(x) = \frac{-9}{x+3}$$

    a. Find equations for the vertical asymptotes, if any, for the function.

    b. Find equations for the horizontal asymptotes, if any, for the function.

    c. Identify four ordered pairs on the graph of the function.

3.9 Rational Inequalities

Rational Inequalities

A _____ is any inequality that can be written in the form:

$$f(x) < 0, f(x) \leq 0, f(x) > 0, \text{ or } f(x) \geq 0,$$

where $f(x)$ is a rational function.

Solving Rational Inequalities: Sign-Test Method

To solve a rational inequality $f(x) < 0$, $f(x) \leq 0$, $f(x) > 0$, or $f(x) \geq 0$, where the rational function $f(x) = \dfrac{p(x)}{q(x)}$ is in reduced form:

Step 1: Find the real _____ of the numerator $p(x)$. These values are the _____ of f.

Step 2: Find the real _____ of the denominator $q(x)$. These values are the locations of the _____ of f.

Step 3: Place the values from Steps 1 and 2 on a number line, splitting it into _____.

Step 4: Within each interval, select a _____ and evaluate f at that number. If the result is positive, then $f(x) > 0$ for all x in the interval. If the result is negative, then $f(x) < 0$ for all x in the interval.

Step 5: Write the _____ , consisting of all of the intervals that satisfy the given inequality. If the inequality is not strict (uses $\leq$ or $\geq$), then the zeros of p are _____ in the solution set as well. The zeros of q are _____ included in the solution set (as they are not in the domain of f).

EXAMPLE 1: Solving Rational Inequalities

Solve the rational inequality $\dfrac{x^2 + 1}{x^2 + 2x - 15} > 0$.

Solution:

CAUTION!

It is very important to write rational inequalities in their proper form before solving them. Attempting to simplify them may cause us to _____ asymptotes. Suppose, for example, we tried to solve the inequality $\dfrac{x}{x+2} < 3$

by multiplying through by $x + 2$, thus clearing the inequality of fractions. The result would be

$$x < 3x + 6,$$

a simple _____ inequality. We can solve this to obtain $x > -3$, or the interval $(-3, \infty)$. But does this really work?

The number $-\dfrac{5}{2}$ is in this interval, but $\dfrac{-\dfrac{5}{2}}{-\dfrac{5}{2}+2} = 5$,

which is not less than 3. Also, note that -4 solves the inequality, but -4 is not in the interval $(-3, \infty)$.

What went wrong? By multiplying through by $x + 2$, we made the assumption that $x + 2$ was _____, since we did not worry about possibly reversing the _____. But we don't know beforehand if $x + 2$ is positive or not, since we are trying to solve for the variable x!

To solve this inequality correctly, we have to write it in the standard form of a rational inequality, as shown in Example 5.

EXAMPLE 2: Rational Inequalities

Solve the rational inequalities $\dfrac{x}{x+2} < 3$ and $\dfrac{x}{x+2} \leq 3$.

Solution

3.9 Exercises

1. Solve the rational inequality. Express your answer in interval notation.

$$6x > \frac{36}{x+5}$$

2. Solve the rational inequality. Express your answer in interval notation.

$$\frac{3}{x^2+2x-8} > \frac{x}{x^2+2x-8}$$

3. Solve the rational inequality. Express your answer in interval notation.

$$\frac{4x}{x+6} < 0$$

4. Solve the rational inequality. Express your answer in interval notation.

$$\frac{x+5}{x} \geq 0$$

5. Solve the rational inequality. Express your answer in interval notation.

$$\frac{-5}{x+3} \geq \frac{-5}{x-3}$$

6. Solve the rational inequality. Express your answer in interval notation.

$$\frac{3}{x+3} \leq \frac{2x}{x+3}$$

7. Solve the rational inequality. Express your answer in interval notation.

$$\frac{5}{x^2 + x - 6} \geq \frac{x}{x^2 + x - 6}$$

8. Solve the rational inequality. Express your answer in interval notation.

$$\frac{x + 3}{x - 4} \leq \frac{x}{x + 4}$$

9. Consider the following rational inequality.

$$\frac{3}{x - 5} \geq \frac{3}{x + 4}$$

  a. Set one side equal to zero and list the interval endpoints (the only points at which the non-zero expression can change sign).

  b. Test each interval to find which ones satisfy the inequality. Express your answer in interval notation.

10. Consider the following rational inequality.

$$\frac{-3}{x^2 + 2x - 15} \leq \frac{x}{x^2 + 2x - 15}$$

  a. Set one side equal to zero and list the interval endpoints (the only points at which the non-zero expression can change sign).

  b. Test each interval to find which ones satisfy the inequality. Express your answer in interval notation.

CHAPTER 4

Exponential and Logarithmic Functions

4.1 Exponential Functions and Their Graphs

Exponential Functions

Let a be a fixed, positive real number not equal to 1. The _____ **function with base a is** the function

$$f\left(x\right) = \text{_____} .$$

Behavior of Exponential Functions

Given a positive real number a not equal to 1, the function $f\left(x\right) = a^x$ is:

- a _____ **function** if $0 < a < 1$, with $f\left(x\right) \to \infty$ as $x \to -\infty$ and $f\left(x\right) \to 0$ as $x \to \infty$.

- an _____ **function** if $a > 1$, with $f\left(x\right) \to 0$ as $x \to -\infty$ and $f\left(x\right) \to \infty$ as $x \to \infty$.

In either case, the point _____ lies on the graph of f, the _____ of f is the set of real numbers, and the _____ of f is the set of positive real numbers.

EXAMPLE 1: Graphing Exponential Functions

Sketch the graphs of the following exponential functions:

a. $f\left(x\right) = 3^x$

b. $g\left(x\right) = \left(\dfrac{1}{2}\right)^x$

Note:

Plugging in $x = -1$, $x = 0$, and $x = 1$ will produce a good idea of the shape of the graph.

Solution

EXAMPLE 2: Graphing Exponential Functions

Sketch the graphs of each of the following functions.

a. $f(x) = \left(\dfrac{1}{2}\right)^{x+3}$

b. $g(x) = -3^x + 1$

c. $h(x) = 2^{-x}$

Note:

For each example, first graph the base function, then apply any transformations.

Solution

Solving Elementary Exponential Equations

To solve an elementary exponential equation,

Step 1: Isolate the _____. Move the exponential containing x to one side of the equation and any constants or other variables in the expression to the other side. Simplify, if necessary.

Step 2: Find a _____ that can be used to rewrite both sides of the equation.

Step 3: Equate the _____, and solve the resulting equation.

EXAMPLE 3: Solving Elementary Exponential Equations

Solve the following exponential equations.

a. $25^x - 125 = 0$

b. $8^{y-1} = \dfrac{1}{2}$

c. $\left(\dfrac{2}{3}\right)^x = \dfrac{9}{4}$

Note:

As always, it is good practice to check your solution in the original equation.

Solution

> **CAUTION!**
>
> Regardless of which method is used, be careful with your negative signs. We know that $-3 \cdot -3 = 9$ and not -9 because a negative times a negative is a positive. However, if we type -3^2 into a calculator, the output is -9. The number that we are squaring is -3, not 3, so we need to type $(-3)^2$ or $(-3)\ ^\wedge\ 2$ into the calculator to get the correct answer, 9.

4.1 Exercises

1. Sketch the graph of the following function.

$$m(x) = \left(\frac{15}{4}\right)^x$$

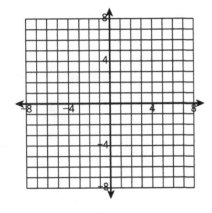

2. Sketch the graph of the following function.

$$r(x) = \left(\frac{1}{2}\right)^x$$

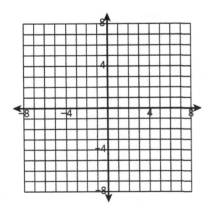

3. Sketch the graph of the following function.

$$f(x) = \left(\frac{15}{4}\right)^{-x}$$

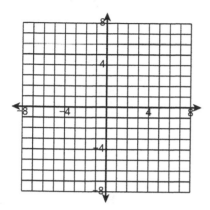

4. Sketch the graph of the following function.

$$p(x) = \frac{1}{3^{1-x}}$$

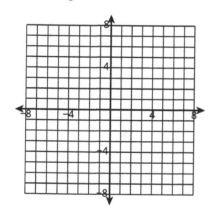

5. Sketch the graph of the following function.

$$g(x) = \left(\frac{1}{5}\right)^{3-x}$$

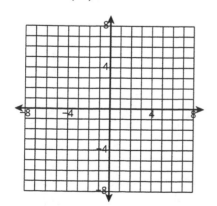

6. Sketch the graph of the following function.

$$q(x) = 1 - \left(\frac{9}{4}\right)^x$$

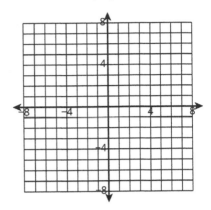

7. Sketch the graph of the following function.

$$q(x) = 3 + \left(\frac{1}{4}\right)^{3-x}$$

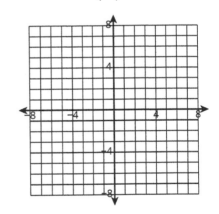

8. Solve the following elementary exponential equation.

$$16^x - 4 = 0$$

9. Solve the following elementary exponential equation.

$$\left(\frac{2}{3}\right)^{3x+2} = \left(\frac{4}{9}\right)^3$$

10. Solve the following elementary exponential equation.

$$6^{3x-4} = \frac{1}{36}$$

4.2 Applications of Exponential Functions

EXAMPLE 1: Population Growth

A biologist is culturing bacteria in a large Petri dish. She begins with 1000 bacteria, and supplies sufficient food so that for the first five hours the bacteria population grows exponentially, doubling every hour.

   a. Find a function that models the population growth of this bacteria culture.

   b. Determine when the population reaches 16,000 bacteria.

   c. Calculate the population two and a half hours after the scientist begins.

Solution

EXAMPLE 2: Radioactive Decay

Determine the base, a, so that the function $A(t) = A_0 a^t$ accurately describes the decay of carbon-14 as a function of t years.

Solution

Compound Interest Formula

An investment of P dollars, compounded n times per year at an annual interest rate of r, has a value after t years of

$$A(t) = \underline{\hspace{4cm}}.$$

EXAMPLE 3: Compound Interest Formula

Sandy invests $10,000 in a savings account earning 4.5% annual interest compounded quarterly. What is the value of her investment after three and a half years?

Solution

EXAMPLE 4: Compound Interest Formula

Nine months after depositing $520.00 in a monthly-compounded savings account, Frank checks his balance and finds the account has $528.84. Being the forgetful type, he can't remember what the annual interest rate for his account is, and sees the bank is advertising a rate of 2.5% for new accounts. Should he close out his existing account and open a new one?

Solution

The Number e

The number e is defined as the value of $\left(1 + \dfrac{1}{m}\right)^m$ as $m \to \infty$.

$$e \approx 2.71828182846$$

In most applications, the approximation $e \approx$ _____ is sufficiently accurate.

Continuous Compounding Formula

An investment of P dollars compounded continuously at an annual interest rate of r has a value after t years of:

$$A\left(t\right) = \underline{\hspace{3cm}}.$$

EXAMPLE 5: Continuous Compounding Formula

If Sandy (last seen in Example 3) has the option of investing her \$10,000 in a continuously compounded account earning 4.5% annual interest, what will be the value of her account in three and a half years?

Solution

4.2 Exercises

1. Tyrell is working in a lab testing bacteria populations. After starting out with a population of 411 bacteria, he observes the change in population and notices that the population doubles every 13 minutes.

    a. Find the equation for the population P in terms of time t in minutes. Round values to three decimal places.

    b. Find the population after 1.5 hours. Round to the nearest bacterium.

2. The half-life of gold-194 is approximately 1.6 days.

    a. Determine a so that $A\left(t\right) = A_0 a^t$ describes the amount of gold-194 left after t days, where A_0 is the amount at time $t = 0$. Round to six decimal places.

b. How much of a 7 gram sample of gold-194 would remain after 4 days? Round to three decimal places.

c. How much of a 7 gram sample of gold-194 would remain after 3 days? Round to three decimal places.

3. The population of a certain inner-city area is estimated to be declining according to the model $P(t) = 254{,}000e^{-0.01t}$, where t is the number of years from the present. What does this model predict the population will be in 7 years? Round to the nearest person.

4. A certain species of fish is to be introduced into a lake, and wildlife experts estimate the population will grow to $P(t) = (408)\, 4^{\frac{t}{2}}$, where t represents the number of years from the time of introduction.

a. What is the quadrupling-time for this population of fish?

b. How long will it take for the population to reach 6528 fish, according to this model?

5. Chester hopes to earn $1200 in interest in 3.4 years time from $48,000 that he has available to invest. To decide if it's feasible to do this by investing in an account that compounds semi-annually, he needs to determine the annual interest rate such an account would have to offer for him to meet his goal. What would the annual rate of interest have to be? Round to two decimal places.

6. Reba has $4900 that she wants to invest in a IRA account for 2.9 years, at which time she plans to close out the account and use the money as a down payment on a car. She finds one local bank offering an annual interest rate of 2.84% compounded semi-annually (Bank 1), and another bank offering an annual interest rate of 2.57% compounded annually (Bank 2). Which bank should she choose?

7. The function $C(t) = C(1 + r)^t$ models the rise in the cost of a product that has a cost of C today, subject to an average yearly inflation rate of r for t years. If the average annual rate of inflation over the next 9 years is assumed to be 1.5%, what will the inflation-adjusted cost of a $31,500 motorcycle be in 9 years? Round to two decimal places.

8. Dolores invests $8200 in a new savings account which earns 5.8% annual interest, compounded continuously. What will be the value of her investment after 4 years? Round to the nearest cent.

9. Monica has recently inherited $7800, which she wants to deposit into a CD account. She has determined that her two best bets are an account that compounds quarterly at an annual rate of 2.8% (Account 1) and an account that compounds annually at an annual rate of 4.2% (Account 2).

   a. Which account would pay Monica more interest?

   b. How much would Monica's balance be from Account 2 over 1.6 years? Round to two decimal places.

10. Statistics show that the fractional part of a pair of flashlight batteries, P, that is still good after t hours of use is given by $P = 4^{-0.02t}$. What fractional part of the batteries is still operating after 150 hours of use? Express your answer as a fraction or as a decimal rounded to six decimal places.

4.3 Logarithmic Functions and Their Graphs

Logarithmic Functions

Let a be a fixed positive real number not equal to 1. The **logarithmic function with base a** is defined to be the _____ of the exponential function with base a, and is denoted _____. In symbols, if $f(x) = a^x$, then $f^{-1}(x) = \log_a x$.

In equation form, the definition of logarithm means that the equations

_____ and _____

are equivalent. Note that a is the base in both equations: either the base of the exponential function or the base of the logarithmic function.

EXAMPLE 1: Exponential and Logarithmic Equations

Use the definition of logarithmic functions to rewrite the following exponential equations as logarithmic equations.

a. $8 = 2^3$

b. $5^4 = 625$

c. $7^x = z$

Then rewrite the following logarithmic equations as exponential equations.

d. $\log_3 9 = 2$

e. $3 = \log_8 512$

f. $y = \log_2 x$

Solution

CAUTION!

While it may appear so on the graph, logarithmic functions do **not** have a _____ asymptote. The reason that logarithmic functions often look like they have a horizontal asymptote is because they are among the slowest growing functions in mathematics! Consider the function $f(x) = \log_2 x$. We have $f(1024) = 10, f(2048) = 11$, and $f(4096) = 12$. While the function may increase its value very _____ as x increases, it never approaches an asymptote.

EXAMPLE 2: Graphing Logarithmic Functions

Sketch the graphs of the following logarithmic functions.

a. $f(x) = \log_3 x$

b. $g(x) = \log_{\frac{1}{2}} x$

Note:

Once again, plotting a few key points will provide a good idea of the shape of the function.

Solution

EXAMPLE 3: Graphing Logarithmic Functions

Sketch the graphs of the following functions.

a. $f(x) = \log_3 (x + 2) + 1$

b. $g(x) = \log_2 (-x - 1)$

c. $h(x) = \log_{\frac{1}{2}} (x) - 2$

Note:

Begin by graphing the base function, then apply any transformations.

Solution

EXAMPLE 4: Logarithmic Expressions

Evaluate the following logarithmic expressions.

a. $\log_5 25$

b. $\log_{17} 1$

c. $\log_{16} 4$

d. $\log_{10}\left(\dfrac{1}{100}\right)$

Note:

We write each of the equivalent exponential equations as a reference.

Solution

EXAMPLE 5: Solving Logarithmic Equations

Solve the following equations involving logarithms.

a. $\log_6 (2x) = -1$

b. $3^{\log_{3x} 2} = 2$

c. $\log_2 8^x = 5$

Solution

Common and Natural Logarithms

- The function $\log_{10} x$ is called the _____ , and is usually written

  _____ .

- The function $\log_e x$ is called the _____ , and is usually written

  _____ .

Properties of Natural Logarithms

$\ln x = y \Leftrightarrow e^y = x$

	Properties	Reasons
1.	$\ln 1 = $ _____	Raise e to the power 0 to get 1.
2.	$\ln e = $ _____	Raise e to the power 1 to get e.
3.	$\ln e^x = $ _____	Raise e to the power x to get e^x.
4.	$e^{\ln x} = $ _____	$\ln x$ is the power to which e must be raised to get x.

EXAMPLE 6: Evaluating Logarithmic Expressions

Evaluate the following logarithmic expressions.

　a. $\ln \sqrt[3]{e}$

　b. $\log 1000$

　c. $\ln (4.78)$

　d. $\log (10.5)$

Solution

4.3 Exercises

1. Write the following equation in logarithmic terms. Do not simplify your answer.

$3.1 = c^3$

2. Write the following logarithmic equation as an exponential equation. Do not simplify your answer.

$4 = \log_{4.4}(V)$

3. Sketch the graph of the following function.

$$q(x) = \log_7(x - 3)$$

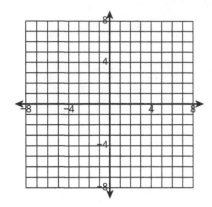

4. Sketch the graph of the following function.

$$m(x) = \log_{\frac{1}{3}}(x) + 4$$

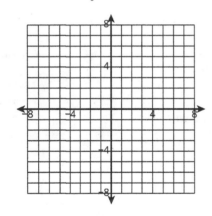

5. Evaluate the following logarithmic expression without the use of a calculator. Write your answer as a fraction reduced to lowest terms.

$\log(0.01)$

6. Evaluate the following logarithmic expression without the use of a calculator. Write your answer as a fraction reduced to lowest terms.

$\log_{17}(289)$

7. Evaluate the following logarithmic expression without the use of a calculator. Write your answer as a fraction reduced to lowest terms.

$\log_7(1)$

8. Evaluate the following logarithmic expression without the use of a calculator. Write your answer as a fraction reduced to lowest terms.

$\log_3\left(\sqrt[6]{81}\right)$

9. Use the elementary properties of logarithms to solve the following equation. Write your answer as a fraction reduced to lowest terms.

$8^{\log_4(x)} = 64$

10. Solve the following logarithmic equation, using a calculator if necessary to evaluate the logarithm. Write your answer as a fraction or round your answer to one decimal place.

$\ln(e^x) = 5.5$

4.4 Applications of Logarithmic Functions

EXAMPLE 1: Continuously Compounded Interest

Anne reads an ad in the paper for a new bank in town. The bank is advertising "continuously compounded savings accounts" in an attempt to attract customers, but fails to mention the annual interest rate. Curious, she goes to the bank and is told by an account agent that if she were to invest, $10,000 in an account, her money would grow to $10,202.01 in one year's time. But, strangely, the agent also refuses to divulge the yearly interest rate. What rate is the bank offering?

Solution

EXAMPLE 2: Solving Exponential Equations

Solve the equation $2^x = 9$.

Solution

Properties of Logarithms

Let a (the logarithmic base) be a positive real number not equal to 1, let x and y be positive real numbers, and let r be any real number.

1. $\log_a (xy) =$ _____   ("the log of a product is the _____ of the logs")

2. $\log_a \left(\dfrac{x}{y} \right) =$ _____   ("the log of a quotient is the _____

  of the logs")

3. $\log_a (x^r) =$ _____   ("the log of something raised to a power is the power _____

  the log")

CAUTION!

Errors in working with logarithms often arise from incorrect recall of the logarithmic properties. The table below highlights some common mistakes.

Incorrect Statements	Correct Statements
$\log_a (x + y) = \log_a x + \log_a y$	$\log_a (xy) = \log_a x + \underline{\hspace{1.5cm}}$
$\log_a (xy) = (\log_a x)(\log_a y)$	$\log_a (xy) = \underline{\hspace{1.5cm}} + \log_a y$
$\dfrac{\log_a x}{\log_a y} = \log_a x - \log_a y$	$\log_a \left(\dfrac{x}{y}\right) = \log_a x - \underline{\hspace{1.5cm}}$
$\dfrac{\log_a x}{\log_a y} = \log_a \left(\dfrac{x}{y}\right)$	$\log_a \left(\dfrac{x}{y}\right) = \underline{\hspace{1.5cm}} - \log_a y$
$\dfrac{\log_a (xz)}{\log_a (yz)} = \dfrac{\log_a x}{\log_a y}$	$\dfrac{\log_a (xz)}{\log_a (yz)} = \dfrac{\log_a x + \log_a z}{\underline{\hspace{2cm}}}$

EXAMPLE 3: Expanding Logarithmic Expressions

Use the properties of logarithms to expand the following expressions as much as possible (that is, decompose the expressions into sums or differences of the simplest possible terms).

a. $\log_4 \left(64x^3 \sqrt{y}\right)$

b. $\log_a \sqrt[3]{\dfrac{xy^2}{z^4}}$

Note:

As long as the base is the $\underline{\hspace{1.5cm}}$ for each term, its value does not affect the use of the properties.

Solution

EXAMPLE 4: Condensing Logarithmic Expressions

Use the properties of logarithms to condense the following expressions as much as possible (that is, rewrite the expressions as a sum or difference of as few logarithms as possible).

a. $2\log_3\left(\dfrac{x}{3}\right) - \log_3\left(\dfrac{1}{y}\right)$

b. $\ln x^2 - \dfrac{1}{2}\ln y + \ln 2$

c. $\log_b 5 + 2\log_b x^{-1}$

> **Note:**
>
> Often, there will be multiple orders in which we can apply the properties to find the final result.

Solution

Change of Base Formula

Let a and b both be positive real numbers, neither of them equal to 1, and let x be a positive real number. Then

$$\log_b x = \underline{\hspace{5cm}}.$$

EXAMPLE 5: Change of Base Formula

Evaluate the following logarithmic expressions, using the base of your choice.

a. $\log_7 15$

b. $\log_{\frac{1}{2}} 3$

Note:

Both the common and natural logarithm work in solving these problems.

Solution

The pH Scale

The **pH** of a solution is defined to be $-\log[H_3O^+]$, where $[H_3O^+]$ is the concentration of _____ in units of moles / liter. Solutions with a pH less than 7 are said to be _____ , while those with a pH greater than 7 are _____ .

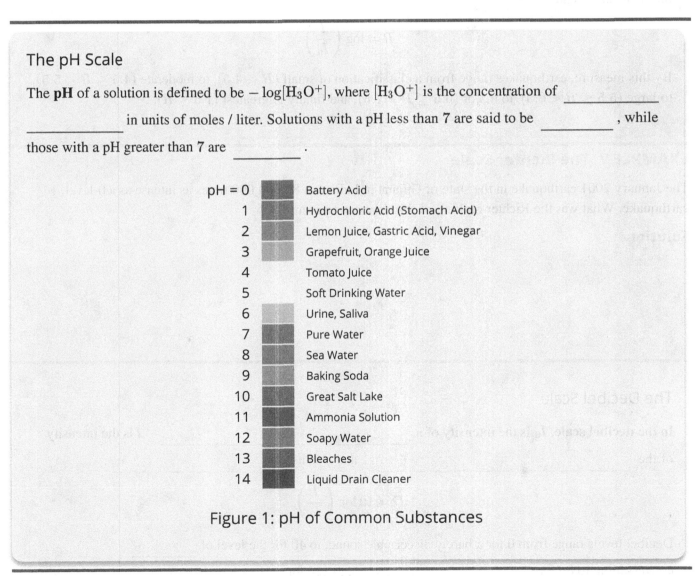

pH = 0		Battery Acid
1		Hydrochloric Acid (Stomach Acid)
2		Lemon Juice, Gastric Acid, Vinegar
3		Grapefruit, Orange Juice
4		Tomato Juice
5		Soft Drinking Water
6		Urine, Saliva
7		Pure Water
8		Sea Water
9		Baking Soda
10		Great Salt Lake
11		Ammonia Solution
12		Soapy Water
13		Bleaches
14		Liquid Drain Cleaner

Figure 1: pH of Common Substances

EXAMPLE 6: The pH Scale

If a sample of orange juice is determined to have a $[H_3O^+]$ concentration of 1.58×10^{-4} moles / liter, what is its pH?

Solution

The Richter Scale

Earthquake intensity is measured on the _____ (named for the American seismologist Charles Richter, 1900–1985). In the formula that follows, _____ is the intensity of a just-discernible earthquake, _____ is the intensity of an earthquake being analyzed, and _____ is its ranking on the Richter scale.

$$R = \log\left(\frac{I}{I_0}\right)$$

By this measure, earthquakes range from a classification of small $(R < 4.5)$, to moderate $(4.5 \leq R < 5.5)$, to large $(5.5 \leq R < 6.5)$, to major $(6.5 \leq R < 7.5)$, and finally to greatest $(7.5 \leq R)$.

EXAMPLE 7: The Richter Scale

The January 2001 earthquake in the state of Gujarat in India was 80,000,000 times as intense as a 0-level earthquake. What was the Richter ranking of this devastating event?

Solution

The Decibel Scale

In the **decibel scale**, I_0 is the intensity of a _____ ,I is the intensity of the _____ , and D is its _____ :

$$D = 10\log\left(\frac{I}{I_0}\right)$$

Decibel levels range from 0 for a barely discernible sound, to 40 for the level of _____ conversation, to 80 for _____ , to 120 for _____ , and finally (as far as humans are concerned) to around 160, at which point the eardrum is likely to _____ .

EXAMPLE 8: The Decibel Scale

Given that $I_0 = 10^{-12}$ watts / meter^2, what is the decibel level of jet airliner's engines at a distance of 45 meters, for which the sound intensity is 50 watts / meter^2?

Solution

4.4 Exercises

1. Use the properties of logarithms to write the following expression as a single term that doesn't contain a logarithm.

   $5^{\log_5(x^3) - 2\log_5(y)}$

2. Use the properties of logarithms to write the following expression as a single term that doesn't contain a logarithm.

   $e^{5\ln\left(\sqrt[9]{14}\right) + 6\ln(x)}$

3. Use the properties of logarithms to write the following expression as a single term that doesn't contain a logarithm.

   $7^{6\log_7(x)}$

4. Use the properties of logarithms to condense the following expression as much as possible, writing the answer as a single term with a coefficient of 1. All exponents should be positive.

   $\ln(15) + 5\ln(x) - 6\ln(y)$

5. Use the properties of logarithms to expand the following expression as much as possible. Simplify any numerical expressions that can be evaluated without a calculator.

$$\log\left(\frac{1.7 \times 10^{11}}{x^{-9}}\right)$$

6. Use the properties of logarithms to expand the following expression as much as possible. Simplify any numerical expressions that can be evaluated without a calculator.

$$\log_6\left(36x^8\right)$$

7. Use the properties of logarithms to expand the following expression as much as possible. Simplify any numerical expressions that can be evaluated without a calculator.

$$\ln\left(\frac{15x^9}{y^6}\right)$$

8. Evaluate the following logarithmic expression. Round off your answer to two decimal places.

$$\log_{\frac{1}{9}}(883)$$

9. If a sample of a certain solution is determined to have a $\left[H_3O^+\right]$ concentration of 5.6×10^{-12} moles/liter, what is its pH? Round off your answer to one decimal place.

10. If an earthquake is 2000000 times as intense as a 0-level earthquake, what is the Richter Scale ranking of this earthquake? Round off your answer to one decimal place.

CHAPTER 5

Mathematics of Finance

5.1 Basics of Personal Finance

Net Income

Net income is equal to total, or gross, income _____ .

Example 1: Determining Federal Income Tax

Jamie graduated from college in 2013 with a degree in accounting. Her first job pays an annual salary of $52,000. How much should Jamie expect to pay in federal income tax for this salary if the federal tax rate is 15%?

Solution

Example 2: Determining Monthly Take Home Pay

Pria just graduated from a liberal arts college and acquired a job as a sociologist. Her yearly salary is $34,500. If the federal tax rate is 15%, the social security tax rate is 6.2%, and the unemployment tax rate is 6%, determine the following.

  a. Pria's yearly taxes.

  b. Pria's monthly take home pay.

Solution

Example 3: Budgeting on a Given Income

In Example 2, Pria had a monthly net income of $2093. Use the steps to create a budget to allocate Pria's monthly income if her expenses consist of the following.

rent:	$600
gas:	$250
utilities:	$150
food/entertainment:	$600
student loans:	$210
	$1810

Solution

List Price

The price of an item as it is listed for public sale is the _____ **price**.

Discount

The **discount** is the _____. This is usually given as a _____ of the list price.

Sale Price

The **sale price**, sometimes called the **net price**, is the actual cost of an item ____ _____ of some kind is applied.

$$\text{sale price} = \text{list price} - \text{discount}$$

Example 5: Finding the Sale Price

Let's begin with a simple example regarding the price of a pair of shoes.

Tennis Star Shoes is having a sale. A pair of tennis shoes has a list price of $129.99. The store is offering a discount of 25% off of the list price. Determine the sale price of the shoes before taxes.

Solution

Percentage Change

The **percentage change** over a period of time can be calculated by the following formula.

percentage change =

Example 7: Computing Percentage Decrease

Freddie bought a tablet PC for $1200. One year later, the same tablet PC costs $900. Determine the percentage decrease for the price of the tablet PC.

Solution

5.1 Exercises

1. Using the information provided in the given table, determine how much monthly income would be necessary to budget in order to cover the expenses of attending a local college for the 9-month academic year. Round your answer to the nearest cent, if necessary.

Estimated Cost of Local College Next Academic Year	
Budget Category	**On-Campus Student**
University Fees	$17,828
Room & Board	$8009
Books & Supplies	$1969
Transportation	$698
Personal	$1817
Health Insurance	$1573
Loan Fees	$168

2. During the last year the value of my car depreciated by 30%. If the value of my car is $25,000 today, then what was the value of my car one year ago? Round your answer to the nearest cent, if necessary .

3. You decide to work fewer hours per week, which results in a 20% decrease in your pay. What percentage increase in pay would you have to receive in order to gain your original salary again? Round your answer to the nearest tenth of a percent.

4. Elijah was recently hired for a job with an annual income of $52,000. Using the federal income rate of 15%, what amount should Elijah expect to pay in federal income tax?

5. Lucy has a job where she has a take-home salary each month of $1700. If Lucy wants to spend no more than 25% of her monthly take-home salary on rent, how much can she afford?

6. A refrigerator with a list price of $2700 will be discounted 25% at the time of purchase. What is the sale price before taxes?

7. At a restaurant, your total bill is $45.75. You wish to give a tip of 18% of the total bill. What is the amount of the tip? Round your answer to the nearest cent.

8. Riley found a receipt for a pair of jeans for $152.97, tax included. If the sales tax rate was 7%, what was the list price of the jeans? Round your answer to the nearest cent.

9. The average price of a certain model of boat in 1995 was $19,500. In 2015, the average price of the boat was $27,300. What is the percentage increase in the average price of the boat?

10. The local sales tax is 8%. If a pair of shoes sells for $145, what is the total cost after tax?

5.2 Simple and Compound Interest

Interest

Interest is the amount charged by a lender for _____ money.

Principal

The **principal** is the sum of money on which _____ is charged.

Interest Rate

The **interest rate** is the amount charged to the borrower expressed as a

_____ .

Annual Percentage Rate (APR)

Annual percentage rate is the yearly interest rate that is charged for borrowing. APR is normally given as

a _____ .

Simple Interest Formula

The amount of interest on a simple interest loan with _____ P, _____

_____ r (written as a decimal), and _____ of t (usually in years) is

calculated with the following formula.

_____

Example 1: Calculating Simple Interest

Determine the interest that is accrued on $5500 for five years at a rate of 8.5%.

Solution

Compound Interest

Compound interest is interest that is computed based on both _____ _____ at each interval. The _____ is the total amount of money A that has been accrued after compounding at an annual percentage rate r based on the initial principal P with n compounding intervals per year for t years. Future value of a compound interest account is calculated with the following formula.

_____

Example 3: Computing Compound Interest

Lilly deposits $12,000 into an account with an annual interest rate of 4.5% compounded monthly. If she leaves the money in the account for 10 years, what will the future value be at the end of this time period?

Solution

Continuous Compound Interest Formula

The future value A of a _____ compound interest account after t years at an annual interest rate of r and an initial amount, or principal, P is calculated with the following formula.

_____

Example 5: Calculating Continuous Compound Interest

Find the future value of $8900 invested at a rate of 2.05% that is compounded continuously over 15 years.

Solution

Annual Percentage Yield (APY)

The **annual percentage yield** (APY) is the effective annual interest rate earned in a given year that accounts for the effects of _____. APY is calculated with the formula

$$APY = \left[\left(1 + \frac{r}{n}\right)^n - 1\right] \cdot 100\%,$$

where r is the _____ and n is the _____

_____ per year.

Example 7: Annual Percentage Yield

Samantha deposits $5000 in an account paying 5% interest per year.

a. Find the APY for Samantha's investment if the interest is compounded monthly.

Solution

Example 9: Calculating Interest on Payday Loans

Assume you wish to borrow $300 for two weeks in the form of a payday loan and the amount of interest you must pay is $25 per $100 borrowed. This means that at the end of two weeks, you owe $375. What is the APR?

Solution

5.2 Exercises

1. A couple deposits \$18,000 into an account earning 4% annual interest for 25 years. Calculate the future value of the investment if the interest is compounded daily. Round your answer to the nearest cent.

2. The First Bank of Lending lists the following APR for loans. Determine the APY, or effective interest rate, for a loan amount that is less than \$20,000. Round your answer to the nearest hundredth, if necessary.

First Bank of Lending Loan APR	
Loan Amount	**APR***
< \$20,000	12.25%
\$20,000–\$99,999	9.99%
> \$99,999	6.75%

* interest rates are compounded quarterly

3. Determine the simple interest earned on \$10,000 after 20 years if the APR is 9%.

4. Determine the simple interest earned on \$22,500 after 15 years if the APR is 11%.

5. Suppose your salary in 2014 is \$70,000. If the annual inflation rate is 5%, what salary do you need to make in 2021 in order for it to keep up with inflation? Round your answer to the nearest cent, if necessary.

6. Determine the simple interest owed on \$860 for 3 years at a rate of 15.5%.

7. Suppose you wish to borrow \$300 for two weeks and the amount of interest you must pay is \$20 per \$100 borrowed. What is the APR at which you are borrowing money?

8. Suppose that $15,000 is deposited for five years at 3% APR. Calculate the interest earned if interest is compounded semiannually. Round your answer to the nearest cent.

9. A payday loan is made for four weeks, where the amount of interest owed per $100 borrowed is $30. If you borrow $900 for four weeks, how much do you owe at the end of the four weeks?

10. Garrett deposits $2000. Determine the APY if there is an APR of 6.5% compounded weekly. Round your answer to the nearest hundredth of a percent, if necessary.

5.3 Annuities: Present and Future Value

Present Value (*PV*)

Present value (*PV*) is the amount of _____ needed now in order to _____

_____ amount. Present value is calculated with the formula

$$ \underline{\hspace{6cm}} $$

where *A* is the future value, *r* is the APR, *n* is the number of compound intervals per year, and *t* is time, or term, in years.

Example 1: Computing Principal Needed for Future Value

State College predicts that in 18 years it will take $150,000 to attend the college for four years. Amber has a substantial amount of cash and wishes to save money for her newborn child's college fund. How much should Amber put aside in an account with an APR of 4%, compounded monthly, in order to have $150,000 in the account in 18 years?

Solution

Annuity

An **annuity** is a _____ made into an account, or taken out of an account, over time.

Annuity Formula for Finding Future Value

The future value (*FV*) of an annuity savings account is calculated with the formula

$$\underline{\hspace{10cm}}$$

where *PMT* is the payment amount that is $\underline{\hspace{8cm}}$, *r* is the APR, *n* is the number of regular payments made each year, and *FV* is the future value after *t* years.

Example 3: Calculating Future Value for an Annuity (Savings Plan)

Use the annuity formula to calculate the future value after six months of monthly $50 payments for an annuity savings plan having an APR of 12%.

Solution

Example 4: Calculating Future Value of an IRA Using the Annuity Formula

At the age of 25, Angela starts an IRA (individual retirement account) to save for retirement. She deposits $200 into the account each month. Based on past performance of similar accounts, she expects to obtain an annual interest rate of 8%.

a. How much money will Angela have saved upon retirement at the age of 65?

b. Compare the future value to the total amount Angela deposited over the time period.

Solution

Annuity Formula for Finding Payment Amounts

The regular payment amount (*PMT*) needed to reach a future goal with an annuity savings account is calculated with the formula

_____

where *FV* is the future value desired, *r* is the APR, *n* is the number of regular payments made each year for *t* years.

Example 5: Finding Monthly Payments in Order to Meet a Goal

Suppose you'd like to save enough money to pay cash for your next computer. The goal is to save an extra $3000 over the next three years. What amount of monthly payments must you make into an account that earns 1.25% interest in order to reach your goal?

Solution

5.3 Exercises

1. Use the formula for present value of money to calculate the amount you need to invest now in one lump sum in order to have $100,000 after 18 years with an APR of 9% compounded quarterly. Round your answer to the nearest cent, if necessary.

2. Attucks College predicts that in 12 years it will take $250,000 to attend the college for four years. Hailey has a substantial amount of cash and wishes to invest a lump sum of money for her child's college fund. How much should Hailey put aside in an account with an APR of 11% compounded monthly in order to have $250,000 in the account in 12 years? Round your answer to the nearest cent, if necessary.

3. Riley starts an IRA (Individual Retirement Account) at the age of 24 to save for retirement. She deposits $350 each month. The IRA has an average annual interest rate of 7%. How much money will she have saved when she retires at the age of 65? Round your answer to the nearest cent, if necessary.

4. Blake starts an IRA (Individual Retirement Account) at the age of 28 to save for retirement. He deposits $350 each month. Upon retirement at the age of 65, his retirement savings is $1,280,208.28. Determine the amount of money Blake deposited over the length of the investment and how much he made in interest upon retirement.

5. Matilda wishes to retire at age 67 with $1,900,000 in her retirement account. When she turns 28, she decides to begin depositing money into an account with an APR of 9%. What is the monthly deposit that Matilda must make in order to reach her goal? Round your answer to the nearest cent, if necessary.

6. Suppose you'd like to save enough money to pay cash for your next car. The goal is to save an extra $24,000 over the next 6 years. What amount must be deposited quarterly into an account that earns 4.3% interest in order to reach your goal? Round your answer to the nearest cent, if necessary.

7. James deposits $1240.52 each quarter into an annuity account for his child's college fund. He wishes to accumulate a future value of $75,000 in 12 years. Assuming an APR of 3.8%, how much of the $75,000 will James ultimately deposit in the account, and how much is interest earned? Round your answers to the nearest cent, if necessary.

8. Kevin deposits a fixed monthly amount into an annuity account for his child's college fund. He wishes to accumulate a future value of $65,000 in 18 years. Assuming an APR of 3.8%, how much of the $65,000 will Kevin ultimately deposit in the account, and how much is interest earned? Round your answers to the nearest cent, if necessary.

9. Kevin deposits a fixed monthly amount into an annuity account for his child's college fund. He wishes to accumulate a future value of $65,000 in 18 years. Assuming an APR of 3.8%, how much of the $65,000 will Kevin ultimately deposit in the account, and how much is interest earned? Round your answers to the nearest cent, if necessary.

10. Suppose you wish to retire at the age of 65 with $70,000 in savings. Determine your monthly payment into an IRA if the APR is 8.5% and you begin making payments at 25 years old. Round your answer to the nearest cent, if necessary.

5.4 Borrowing Money

Grace Period

A **grace period** is a period of time in which _____ on a debt.

Number of Fixed Payments Required to Pay Off Credit Card Debt

The _____ **R required to pay off a credit card debt** is calculated with the formula

$$R = \frac{-\log\left[1 - \frac{r}{n}\left(\frac{A}{PMT}\right)\right]}{\log\left(1 + \frac{r}{n}\right)}$$

where n is the number of payments made per year for a loan of amount A, with interest rate r and monthly payment PMT.

Example 1: Paying Off Credit Card Debt with a Fixed Payment

Assume you want to buy a new computer that costs $2200 using a credit card that has an APR of 19.99%.

a. How long will it take you to pay off the computer if you make regular monthly payments of $40?

b. How much will you pay in the long run for the computer if you make monthly payments of $40?

Solution

Fixed Installment Loans

Fixed installment loans have a _____ and are paid off with _____ over a specified amount of time.

Down Payment

A **down payment** is a _____ towards the total purchase price of the goods or service.

Monthly Payment Formula for Fixed Installment Loans

The amount of a **monthly payment** (*PMT*) **on a fixed installment loan** is calculated with the formula

_____

where P is the principal amount of money borrowed, r is the APR, n is the number of payments per year, and t is the number of years of the loan.

Example 3: Determining Best New Car Incentive

Dean gets to choose from one of the new car incentives when he purchases his car next week. He can either choose 0.9% APR financing for 60 months or $1500 cash back with a 3.75% APR over 48 months. Compare the two incentives that Dean has to choose from if the new car he wishes to buy is $27,465 and he has saved a down payment of $5000.

Solution

Example 4: Calculating Monthly Mortgage Payments

William and Helen are preparing to buy a new home in the Midwestern United States. They have saved $21,300 for a down payment. The total price of the house is $131,800, including taxes and fees. Find their monthly mortgage payment if the fixed interest rate is 3.52% for a 30-year loan.

Solution

Maximum Purchase Price Formula

The **maximum purchase price** for a house can be found by

$$\text{maximum purchase price} = $$

where *PMT* is the monthly mortgage payment you wish to make, r is the APR, n is the number of payments per year, and t is the number of years of the loan.

Example 5: How Much House Can You Afford?

Suppose you have recently graduated from college and want to purchase a house. Your take-home pay is $3220 per month and you wish to stay within the recommended guidelines for mortgage amounts by only spending $\frac{1}{4}$ of your take-home pay on a house payment. You have $15,300 saved for a down payment. With your good credit and the down payment you can get an APR from your bank of 3.37% compounded monthly.

a. What is the total cost of a house you could afford with a 15–year mortgage?

Solution

5.4 Exercises

1. Suppose you want to purchase a house. Your take-home pay is $2960 per month, and you wish to stay within the recommended guidelines for mortgage amounts by only spending $\frac{1}{4}$ of your take-home pay on a house payment. You have $16,000 saved for a down payment and you can get an APR from your bank of 4.1%, compounded monthly. What is the total cost of a house you could afford with a 30-year mortgage? Round your answer to the nearest cent, if necessary.

2. Easton has two options for buying a car. Option A is 1.1% APR financing over 60 months and Option B is 5.6% APR over 60 months with $1500 cash back, which he would use as part of the down payment. The price of the car is $26,009 and Easton has saved $2600 for the down payment. Find the total amount Easton will spend on the car for each option if he plans to make monthly payments. Round your answers to the nearest cent, if necessary.

3. You have a credit card that has a balance of $5810 at an APR of 17.99%. You plan to pay $400 each month in an effort to clear the debt quickly. How many months will it take you to pay off the balance?

4. Lynette bought a new car for $27,000. She paid a 20% down payment and financed the remaining balance for 60 months with an APR of 4.2%. Assuming she made monthly payments, determine the total cost of Lynette's car. Round your answer to the nearest cent, if necessary.

5. Carter bought a new car and financed $29,000 to make the purchase. He financed the car for 60 months with an APR of 5.5%. Assuming he made monthly payments, determine the total interest Carter paid over the life of the loan. Round your answer to the nearest cent, if necessary.

6. Michelle bought a new car for $28,000. She paid a 20% down payment and financed the remaining balance for 60 months with an APR of 5.5%. Determine the monthly payment that Michelle pays. Round your answer to the nearest cent, if necessary.

7. Jenna bought a new car for $35,000. She paid a 10% down payment and financed the remaining balance for 48 months with an APR of 4.5%. Assuming she makes monthly payments, determine the total interest Jenna pays over the life of the loan. Round your answer to the nearest cent, if necessary.

8. Teagan wants to buy a new generator. The generator costs $2400. Teagan decides to finance the generator for 24 months at an APR of 11.5%. Determine Teagan's monthly payment. Round your answer to the nearest cent, if necessary.

9. A new school building was recently built in the area. The entire cost of the project was $20,000,000. The city has put the project on a 20-year loan with an APR of 3.3%. There are 15,000 families that will be responsible for making monthly payments towards the loan. Determine the total amount that each family should be required to pay each year to cover the cost of the new school building. Round your answer to the nearest cent, if necessary.

10. Consider a credit card with a balance of $7500 and an APR of 17.99%. In order to pay off the balance in 1 year, what monthly payment would you need to make? Round your answer to the nearest cent, if necessary.

CHAPTER 6

Systems of Linear Equations; Matrices

6.1 Solving Systems of Linear Equations by Substitution and Elimination

Solutions to Linear Systems of Equations

- A linear system of equations with no solution is called _____ .

- A linear system of equations that has at least one solution is called _____ .

- Any linear system with more than one solution must have an infinite number of solutions and is called _____ .

EXAMPLE 1: Solving a System of Equations by Substitution

Use the method of substitution to solve the system $\begin{cases} 2x - y = 1 \\ x + y = 5 \end{cases}$

Note:

Solving for a variable with a coefficient of 1 will make the process of substitution a bit easier.

Solution

EXAMPLE 2: Solving a System of Equations by Substitution

Use the method of substitution to solve the system $\begin{cases} -2x + 6y = 6 \\ x + 3 = 3y \end{cases}$

Solution

EXAMPLE 3: Solving a System of Equations by Elimination

Use the method of elimination to solve the system $\begin{cases} 5x + 3y = -7 \\ 7x - 6y = -20 \end{cases}$

Solution

EXAMPLE 4: Solving a System of Equations by Elimination

Use the method of elimination to solve the system $\begin{cases} 2x - 3y = 3 \\ 3x - \dfrac{9}{2}y = 5 \end{cases}$

Solution

EXAMPLE 5: Solving a System of Equations by Elimination

Solve the system $\begin{cases} 2x + y + z = 6 \\ 2x + 3y - z = -2 \\ -3x + 2y - z = 5 \end{cases}$

Solution

EXAMPLE 8: Determining Ages

If the ages of three girls, Xenia, Yolanda, and Zsa Zsa, are added, the result is 30. The sum of Xenia's and Yolanda's ages is Zsa Zsa's age, while Xenia's age subtracted from Yolanda's is half of Zsa Zsa's age a year ago. How old is each girl?

Solution

6.1 Exercises

1. Use the method of substitution to solve the following system of equations. If the system is dependent, express the solution set in terms of one of the variables. Leave all fractional answers in fraction form.

$$\begin{cases} -y = 31 \\ -7x - y = -39 \end{cases}$$

2. Use the method of substitution to solve the following system of equations. If the system is dependent, express the solution set in terms of one of the variables. Leave all fractional answers in fraction form.

$$\begin{cases} -2x + 5y = 13 \\ 4x - 10y = -26 \end{cases}$$

3. Use the method of elimination to solve the following system of equations. If the system is dependent, express the solution set in terms of one of the variables. Leave all fractional answers in fraction form.

$$\begin{cases} -5x - 6y = 34 \\ x + 4y = -18 \end{cases}$$

4. Use the method of elimination to solve the following system of equations. If the system is dependent, express the solution set in terms of one of the variables. Leave all fractional answers in fraction form.

$$\begin{cases} -3x + 3y = -24 \\ -6x + 6y = -48 \end{cases}$$

5. Determine whether the following system of linear equations has no solution, only one solution, or infinitely many solutions. If the system has only one solution, find that solution.

$$\begin{cases} -2x - 2y - 6z = -2 \\ -4x + 4y - 4z = -4 \\ x - 6y - 4z = 1 \end{cases}$$

6. Determine whether the following system of linear equations has no solution, only one solution, or infinitely many solutions. If the system has only one solution, find that solution.

$$\begin{cases} 4x + 7y + 4z = 2 \\ -3x + 6y + 4z = 1 \\ -6x + 4y + 3z = 1 \end{cases}$$

7. Determine whether the following system of linear equations has no solution, only one solution, or infinitely many solutions. If the system has only one solution, find that solution.

$$\begin{cases} -6x + 4y - 2z = -6 \\ x - 3y + z = 5 \\ -3x - 5y + z = 1 \end{cases}$$

8. Use any convenient method to determine whether the following system of equations is consistent, inconsistent, or dependent.

$$\begin{cases} 2x - 6y = 28 \\ -8x + 24y = -114 \end{cases}$$

9. A metallurgist has one alloy containing 23% aluminum and another containing 52% aluminum. How many pounds of each alloy must he use to make 48 pounds of a third alloy containing 28% aluminum? (Round off the answers to the nearest hundredth.)

10. Use a system of equations to solve the following problem.

The sum of three integers is 182. The sum of the first and second integers exceeds the third by 88. The third integer is 16 less than the first. Find the three integers.

6.2 Matrix Notation and Gauss-Jordan Elimination

Matrices and Matrix Notation

A _____ is a rectangular array of numbers, called _____ or **entries** of the matrix. As the numbers are in a rectangular array, they naturally form **rows** and **columns**. It is often important to determine the size of a given matrix; we say that a matrix with m _____ and n _____ is an $m \times n$ matrix (read "m by n"), or of _____ $m \times n$. By convention, the number of rows is always stated first.

$$\text{rows} \begin{bmatrix} 4 & 0 & 1 & -3 \\ 12 & -5 & -5 & 8 \\ 0 & -7 & 25 & 2 \end{bmatrix}$$

columns

This matrix has _____ rows and _____ columns, so it is a _____ matrix.

Matrices are often labeled with _____ letters. The same letter in lower-case, with a pair of subscripts attached, is usually used to refer to its individual _____. For instance, if A is a matrix, a_{ij} refers to the element in the ith row and the jth column.

If we call the above matrix A, then we have $a_{23} = -5$ and $a_{32} = -7$.

EXAMPLE 1: Matrices and Matrix Notation

Given the matrix $A = \begin{bmatrix} -27 & 0 & 1 \\ 5 & -\pi & 13 \end{bmatrix}$, determine:

a. The order of A.

b. The value of a_{13}.

c. The value of a_{21}.

Solution

Standard Form of a System of Linear Equations

A linear system of equations is in **standard form** when each equation has been simplified with its _____ on the left-hand side and its _____ term on the right-hand side. Each equation should have its variable terms listed in the same order.

For example, the system $\begin{cases} 3x + 4 = 7y \\ -2x + 8y = 18 \end{cases}$ is written in standard form as $\begin{cases} 3x - 7y = -4 \\ -x + 4y = 9 \end{cases}$

Augmented Matrices

Given a linear system of equations written in standard form, the _____ of that system is a matrix consisting of the coefficients of the variables listed in their relative positions with an adjoined column of the constants of the system. The matrix of coefficients and the column of constants are customarily separated by a _____ .

For example, the augmented matrix for the system $\begin{cases} 3x - 7y = -4 \\ -x + 4y = 9 \end{cases}$ is $\left[\begin{array}{cc|c} 3 & -7 & -4 \\ -1 & 4 & 9 \end{array}\right]$

The augmented matrix will have as many rows as there are equations in the system, and one more column than there are variables.

EXAMPLE 2: Augmented Matrices

Construct the augmented matrix for the linear system $\begin{cases} \dfrac{2x - 6y}{2} = 3 - z \\ z - x + 5y = 12 \\ x + 3y - 2 = 2z \end{cases}$

Solution

Row Echelon Form

A matrix is in **row echelon form** if:

1. The first nonzero entry in each _____ is 1. We call this a _____ **1.**

2. Every entry _____ a leading 1 is 0, and each leading 1 appears farther to the _____ than the leading 1's in the rows above it.

3. All rows consisting entirely of 0's (if there are any) appear at the _____ .

EXAMPLE 3: Row Echelon Form

Determine if the following matrices are in row echelon form. If not, explain why not.

a. $\begin{bmatrix} 2 & 0 & 13 & | & -1 \\ 0 & 5 & 1 & | & 1 \\ 0 & 0 & 1 & | & -4 \end{bmatrix}$

b. $\begin{bmatrix} 1 & 2 & -8 & | & 0 \\ 0 & 1 & 1 & | & 3 \\ 0 & 0 & 1 & | & 10 \end{bmatrix}$

c. $\begin{bmatrix} 1 & 0 & 4 & | & -2 \\ 1 & 2 & 0 & | & 7 \\ 1 & -5 & -3 & | & 6 \end{bmatrix}$

d. $\begin{bmatrix} 1 & -3 & 1 & | & 6 \\ 0 & 0 & 0 & | & 0 \\ 0 & 1 & -8 & | & -10 \end{bmatrix}$

Solution

Elementary Row Operations

Let A be an augmented matrix corresponding to a system of equations. Each of the following operations on A results in the augmented matrix of an equivalent system. In the notation, R_i refers to row i of the matrix A.

1. Rows i and j can be _____ . (Denoted as $R_i \leftrightarrow R_j$)

2. Each entry in row i can be _____ by a nonzero constant c. (Denoted as cR_i)

3. Row j can be replaced with the _____ of itself and a constant multiple of row i. (Denoted as $cR_i + R_j$)

Typically, we write that an operation has been performed by connecting the original matrix to the new matrix with an _____ , writing the type of operation above it. For example, the matrix from Example 3d can be placed in row echelon form as follows.

$$\begin{bmatrix} 1 & -3 & 1 & 6 \\ 0 & 0 & 0 & 0 \\ 0 & 1 & -8 & -10 \end{bmatrix} \xrightarrow{R_2 \leftrightarrow R_3} \begin{bmatrix} 1 & -3 & 1 & 6 \\ 0 & 1 & -8 & -10 \\ 0 & 0 & 0 & 0 \end{bmatrix}$$

EXAMPLE 4: Elementary Row Operations

Perform the indicated elementary row operation.

$$\begin{bmatrix} 1 & 2 & 3 & 0 \\ 5 & 4 & 1 & -7 \\ 6 & 1 & 8 & -9 \end{bmatrix} \xrightarrow{-5R_1 + R_2} ?$$

Solution

EXAMPLE 5: Gaussian Elimination

Use Gaussian elimination to solve the system $\begin{cases} -2x + y - 5z = -6 \\ x + 2y - z = -8 \\ 3x - y + 2z = 2 \end{cases}$

Solution

Reduced Row Echelon Form

A matrix is said to be in **reduced row echelon form** if:

1. It is in _____ form.

2. Each entry _____ a leading 1 is also 0.

EXAMPLE 6: Reduced Row Echelon Form

Use Gauss-Jordan elimination to solve the system $\begin{cases} x - 2y + 3z = -5 \\ 2x + 3y - z = 1 \\ -x - 5y + 4z = -6 \end{cases}$

Solution

6.2 Exercises

1. Fill in the blank by performing the indicated elementary row operation(s).

$$\left[\begin{array}{cc|c} 4 & 1 & -1 \\ 1 & 9 & 6 \end{array}\right] \xrightarrow{-4R_2+R_1} \underline{\quad ? \quad}$$

2. Fill in the blank by performing the indicated elementary row operation(s).

$$\left[\begin{array}{ccc|c} 1 & 8 & -2 & -4 \\ -4 & -4 & 2 & 8 \\ -2 & 12 & -8 & 2 \end{array}\right] \xrightarrow{4R_1+R_2,\,2R_1+R_3} \underline{\quad ? \quad}$$

3. Consider the following matrix:

Let $D = \left[\begin{array}{ccc} 8 & 2 & -14 \\ -3 & -2 & 0 \end{array}\right]$

   a. Determine the order of D.

   b. Determine the value of d_{22}, if possible. If the element indicated is not in the matrix, state "None".

   c. Determine the value of d_{13}, if possible. If the element indicated is not in the matrix, state "None".

4. Consider the following matrix:

$$\text{Let } C = \begin{bmatrix} -6 & 15 \\ 8 & -9 \\ -13 & 3 \end{bmatrix}$$

a. Determine the order of C.

b. Determine the value of c_{32}, if possible. If the element indicated is not in the matrix, state "None".

c. Determine the value of c_{31}, if possible. If the element indicated is not in the matrix, state "None".

5. Construct the augmented matrix that corresponds to the following system of equations.

$$\begin{cases} \dfrac{2x + 5}{7} - \dfrac{z}{6} = \dfrac{7y}{6} \\ x - 7(y + 3z) = y + 5 \\ 8x + 7 - 2z = -5y \end{cases}$$

6. Use Gaussian elimination and back-substitution to solve the following system of equations. If there is a solution, write your answer in the format (x, y).

$$\begin{cases} 6x - 6y = -18 \\ -4x + 2y = 16 \end{cases}$$

7. Use Gaussian elimination and back-substitution to solve the following system of equations. If there is a solution, write your answer in the format (x, y).

$$\begin{cases} -3x - 4y = -11 \\ -9x - 12y = -35 \end{cases}$$

8. Use Gauss-Jordan elimination to solve the following system of equations. If there is a solution, write your answer in the format (x, y, z).

$$\begin{cases} -3x + 2y + z = -19 \\ x + 7y - 5z = -26 \\ 4x + 5y - 6z = -7 \end{cases}$$

9. Use Gauss-Jordan elimination to solve the following system of equations. If there is a solution, write your answer in the format (x, y, z).

$$\begin{cases} 3x - 3y + 15z = -55 \\ 2x + 2y - 3z = 13 \\ x - y + 5z = -19 \end{cases}$$

10. Use Gauss-Jordan elimination to solve the following system of equations. If there is a solution, write your answer in the format (x, y).

$$\begin{cases} 7x + y = -23 \\ 5x + 7y = -29 \end{cases}$$

6.3 Determinants and Cramer's Rule

Determinant of a 2×2 Matrix

The _____ of the matrix $A = \begin{bmatrix} a_{11} & a_{12} \\ a_{21} & a_{22} \end{bmatrix}$, denoted $|A|$, is

_____ .

CAUTION!

If A is a matrix, $|A|$ stands for the determinant of A, not the _____ of A. In fact, the concept of absolute value does not apply to matrices.

EXAMPLE 1: Determinant of a 2×2 Matrix

Evaluate the determinants of each of the following matrices:

a. $A = \begin{bmatrix} -2 & 3 \\ -1 & 3 \end{bmatrix}$

b. $B = \begin{bmatrix} 2 & -3 \\ -4 & 6 \end{bmatrix}$

Solution

Minors and Cofactors

Let A be an $n \times n$ matrix, and let i and j be numbers between 1 and n, so that a_{ij} is an element of A.

- The **minor** of the element a_{ij} is the determinant of the $(n-1) \times (n-1)$ matrix formed from A by deleting its ith _____ and j^{th} _____.

- The **cofactor** of the element a_{ij} is $(-1)^{i+j}$ times the minor of a_{ij}. Thus, if $i+j$ is _____, the cofactor of a_{ij} is the _____ as the minor of a_{ij}; if $i+j$ is _____, the cofactor is the _____ of the minor.

EXAMPLE 2: Minors and Cofactors

For the matrix $A = \begin{bmatrix} -5 & 3 & 2 \\ 1 & 0 & -1 \\ -3 & 1 & 0 \end{bmatrix}$,

a. Evaluate the minor of a_{12}.

b. Evaluate the cofactor of a_{23}.

Solution

Determinant of an $n \times n$ Matrix

Evaluation of an $n \times n$ determinant is accomplished by _____ along a fixed row or column. The result does not depend on which row or column is chosen.

- To expand along the ith row, each element of that row _____ its cofactor, and the n products are then _____.

- To expand along the jth column, each element of that column is multiplied by its _____, and the n products are then added.

For example, if we expand along the first column of a 3×3 matrix, we get the following. Note the minus sign in front of a_{21}.

$$\begin{vmatrix} a_{11} & a_{12} & a_{13} \\ a_{21} & a_{22} & a_{23} \\ a_{31} & a_{32} & a_{33} \end{vmatrix} = a_{11} \begin{vmatrix} a_{22} & a_{23} \\ a_{32} & a_{33} \end{vmatrix} - a_{21} \begin{vmatrix} a_{12} & a_{13} \\ a_{32} & a_{33} \end{vmatrix} + a_{31} \begin{vmatrix} a_{12} & a_{13} \\ a_{22} & a_{23} \end{vmatrix}$$

We could expand along a row or a different column in a similar manner.

EXAMPLE 3: Determinant of an $n \times n$ Matrix

Evaluate the determinant of the matrix $A = \begin{bmatrix} -1 & 3 & 2 \\ -2 & 0 & 0 \\ 4 & 1 & 5 \end{bmatrix}$.

Note

Minimize the number of computations by choosing which row or column to expand along carefully.

Solution

Properties of Determinants

1. A _____ can be factored out of each of the terms in a given row or column when computing determinants. For example, $\begin{vmatrix} 2 & -1 \\ 15 & 5 \end{vmatrix} = 5 \begin{vmatrix} 2 & -1 \\ 3 & 1 \end{vmatrix}$ and $\begin{vmatrix} 4 & 7 \\ 12 & 9 \end{vmatrix} = 4 \begin{vmatrix} 1 & 7 \\ 3 & 9 \end{vmatrix}$.

2. _____ two rows or two columns changes the determinant by a factor of -1. For example, $\begin{vmatrix} 2 & -1 \\ 15 & 5 \end{vmatrix} = - \begin{vmatrix} 15 & 5 \\ 2 & -1 \end{vmatrix}$ and $\begin{vmatrix} 3 & -2 \\ 7 & 1 \end{vmatrix} = - \begin{vmatrix} -2 & 3 \\ 1 & 7 \end{vmatrix}$.

3. The determinant is _____ by adding a multiple of one row (or column) to another row (or column). For example, $\begin{vmatrix} 3 & -2 \\ 1 & -1 \end{vmatrix} \overset{-3R_2 + R_1}{=\!=} \begin{vmatrix} 0 & 1 \\ 1 & -1 \end{vmatrix}$.

EXAMPLE 4: Properties of Determinants

Evaluate the determinant of the matrix $B = \begin{bmatrix} 4 & -2 & 3 & 0 \\ 2 & 1 & -1 & 3 \\ 3 & 0 & 1 & 1 \\ 2 & -2 & 0 & 0 \end{bmatrix}$.

Solution

Cramer's Rule for Two-Variable, Two-Equation Linear Systems

The solution of a two-variable, two-equation linear system $\begin{cases} ax + by = e \\ cx + dy = f \end{cases}$, is given by

$$x = \frac{D_x}{D} \text{ and } y = \frac{D_y}{D}$$

where D is the determinant of the coefficient matrix $\begin{vmatrix} a & b \\ c & d \end{vmatrix}$, D_x is the determinant of the matrix formed by

replacing the column of x-coefficients with the column of _____ $\begin{vmatrix} e & b \\ f & d \end{vmatrix}$, and D_y

is the determinant of the matrix formed by replacing the column of y-coefficients with the column of

_____ $\begin{vmatrix} a & e \\ c & f \end{vmatrix}$.

CAUTION!

Whenever a fraction appears in our work, we need to ask if the expression in the denominator can ever be zero, and what it means if this happens. In Cramer's Rule, the determinant D can equal 0, which prevents us from using the given formulas.

If $D = 0$, the system is either _____ or _____. If both D_x and D_y are

also zero, the system is _____. If at least one of D_x and D_y is nonzero, the system has

_____.

EXAMPLE 5: Cramer's Rule

Use Cramer's rule to solve the following systems.

a. $\begin{cases} 4x - 5y = 3 \\ -3x + 7y = 1 \end{cases}$

b. $\begin{cases} -x + 2y = -1 \\ 3x - 6y = 3 \end{cases}$

Solution

Cramer's Rule

A linear system of n equations in the n variables $x_1, x_2, \ldots, x_n$ can be written in the form

$$\begin{cases} a_{11}x_1 + a_{12}x_2 + \ldots + a_{1n}x_n = b_1 \\ a_{21}x_1 + a_{22}x_2 + \ldots + a_{2n}x_n = b_2 \\ \vdots \\ a_{n1}x_1 + a_{n2}x_2 + \ldots + a_{nn}x_n = b_n \end{cases}$$

The solution of the system is given by the formulas $x_1 = \dfrac{D_{x_1}}{D}, x_2 = \dfrac{D_{x_2}}{D}, \ldots, x_n = \underline{\hspace{2cm}}$, where D

is the determinant of the $\underline{\hspace{5cm}}$ and D_{x_i} is the determinant of the same matrix

with the ith column replaced by the column of $\underline{\hspace{3cm}} b_1, b_2, \ldots, b_n$.

If $D = 0$ and if each $D_{x_i} = 0$ as well, the system is dependent and has an $\underline{\hspace{3cm}}$

of solutions. If $D = 0$ and at least one of the D_{x_i} is nonzero, the system has $\underline{\hspace{3cm}}$.

EXAMPLE 6: Cramer's Rule

Use Cramer's rule to solve the system $\begin{cases} 3x - 2y - 2z = -1 \\ 3y + z = -7 \\ x + y + 2z = 0 \end{cases}$

Solution

6.3 Exercises

1. Evaluate the determinant of the matrix.

$$A = \begin{bmatrix} 28 & -44 \\ 14 & -22 \end{bmatrix}$$

2. Determine the minor of a_{12} in the matrix.

$$A = \begin{bmatrix} 2 & -3 & 0 \\ 8 & 6 & 1 \\ 6 & 8 & -4 \end{bmatrix}$$

3. Determine the cofactor of a_{13} in the matrix.

$$A = \begin{bmatrix} 5 & -8 & 5 \\ 3 & 5 & 6 \\ 1 & 1 & -6 \end{bmatrix}$$

4. Determine the cofactor of a_{32} in the matrix.

$$A = \begin{bmatrix} 2 & -4 & 6 \\ 9 & 3 & 1 \\ 7 & 5 & -3 \end{bmatrix}$$

5. Evaluate the determinant of the matrix $A = \begin{bmatrix} 5 & 0 & 4 \\ -8 & 4 & 0 \\ 6 & -4 & 4 \end{bmatrix}$. Minimize the required number of computations by carefully choosing a row or column to expand along, and use the properties of determinants to simplify the process.

6. Evaluate the determinant of the matrix $A = \begin{bmatrix} -2 & -1 & -2 & 0 \\ 2 & 2 & 1 & 6 \\ 4 & 0 & 0 & 1 \\ 2 & 0 & 1 & 0 \end{bmatrix}$. Minimize the required number of

computations by carefully choosing a row or column to expand along, and use the properties of determinants to simplify the process.

7. Use Cramer's rule to solve the system $\begin{cases} x - 9y = 1 \\ 9x + 12y = -84 \end{cases}$. If there is a solution, write your answer in the

format (x, y).

8. Use Cramer's rule to solve the system $\begin{cases} -6x + 3y = 6 \\ 2x - y = -2 \end{cases}$. If there is a solution, write your answer in the

format (x, y).

9. Use Cramer's rule to solve the system $\begin{cases} x + 8y = -14 \\ y + 10z = 9 \\ 24x + 6z = -138 \end{cases}$. If there is a solution, write your answer in the

format (x, y, z).

10. Use Cramer's rule to solve the system $\begin{cases} 16w - 12x - 20y = 56 \\ w - 3x + y = -14 \\ x + z = 70 \\ y + z = 0 \end{cases}$. If there is a solution, write your answer

in the format (w, x, y, z).

6.4 Basic Matrix Operations

Matrix Addition

Two matrices A and B can be added to form the new matrix $A + B$ only if A and B are of the

_____ . The addition is performed by adding _____

of the two matrices together; that is, the element in the ith row and the jth column of $A + B$ is given by $a_{ij} + b_{ij}$.

CAUTION!

It is very important to note the restriction on the order of the two matrices in the above definition; a matrix with m rows and n columns can only be added to another matrix with _____ and

_____ . In all aspects of matrix algebra, the orders of the matrices involved must be considered.

Matrix Equality

Two matrices A and B are _____ , denoted $A = B$, if they are of the _____ and all corresponding entries of A and B are _____ .

EXAMPLE 1: Matrix Addition

Perform the indicated addition, if possible.

a. $\begin{bmatrix} -3 & 2 \\ 0 & -5 \\ 11 & -9 \end{bmatrix} + \begin{bmatrix} 3 & 17 \\ 5 & 4 \\ -10 & 4 \end{bmatrix}$

b. $\begin{bmatrix} 2 & -5 \\ 1 & 0 \\ 0 & 3 \\ -7 & 10 \end{bmatrix} + \begin{bmatrix} 2 & 1 & 0 & -7 \\ -5 & 0 & 3 & 10 \end{bmatrix}$

Solution

EXAMPLE 2: Matrix Equations

Determine the values of the variables that will make each of the following statements true.

a. $\begin{bmatrix} -3 & a & b \\ -2 & a+b & 5 \end{bmatrix} = \begin{bmatrix} c & 3 & 7 \\ -2 & d & 5 \end{bmatrix}$

b. $\begin{bmatrix} 3x \\ 4 \end{bmatrix} + \begin{bmatrix} -y \\ 2x \end{bmatrix} = \begin{bmatrix} 13 \\ 7y \end{bmatrix}$

Solution

Scalar Multiplication

If A is an $m \times n$ matrix and c is a scalar, cA stands for the $m \times n$ matrix for which each entry is

_____ the corresponding entry of A. In other words, the entry in the ith row and j^{th} column of

cA is ca_{ij} .

EXAMPLE 3: Scalar Multiplication

Given the matrices $A = \begin{bmatrix} -1 & 6 & 2 \\ -8 & 0 & 1 \end{bmatrix}$ and $B = \begin{bmatrix} 0 & -3 & 4 \\ 1 & -2 & 6 \end{bmatrix}$, write $-3A + 2B$ as a single matrix.

Solution

Matrix Subtraction

Let A and B be two matrices of the same order. The difference $A - B$ is defined by

$$A - B = A + (-B).$$

EXAMPLE 4: Matrix Subtraction

Perform the indicated subtraction: $\begin{bmatrix} 3 & -5 & 2 \end{bmatrix} - \begin{bmatrix} -2 & -5 & 3 \end{bmatrix}$.

Solution

Matrix Multiplication

Two matrices A and B can be multiplied together, resulting in a new matrix denoted AB, only if the number of _____ of A (the matrix on the left) is the same as the number of _____ of B (the matrix on the right). Thus, if A is of order $m \times n$, the product AB is only defined if B is of order $n \times p$. The order of AB will be $m \times p$.

If we let c_{ij} denote the entry in the ith row and j^{th} column of AB, c_{ij} is obtained from the ith row of A and the jth column of B by the formula

$$c_{ij} = a_{i1}b_{1j} + a_{i2}b_{2j} + \ldots + a_{in}b_{nj}.$$

In words, c_{ij} equals the product of the _____ of row i of matrix A and the _____ of column j of matrix B, plus the product of the _____ of row i and the _____ of column j, and so on.

CAUTION!

Unlike numerical multiplication, matrix multiplication is not _____. That is, given two matrices A and B, AB *in general* is not equal to BA. As an illustration of this fact, suppose A is a 3×4 matrix and B is a 4×2 matrix. Then AB is defined (and is of order 3×2), but BA doesn't even exist. Even when both AB and BA are defined, they are generally not equal.

EXAMPLE 6: Matrix Multiplication

Given the matrices $A = \begin{bmatrix} 2 & -3 \\ 4 & -1 \\ 1 & 0 \end{bmatrix}$ and $B = \begin{bmatrix} 5 & 0 & -2 \\ -4 & 1 & 3 \end{bmatrix}$, find the following products.

a. AB

b. BA

Solution

6.4 Exercises

1. Given the following matrices, if possible, determine $A + B + C$. If not, state "Not Possible".

$$A = \begin{bmatrix} -5 & 1 \\ 4 & 1 \\ 4 & -3 \end{bmatrix} \qquad B = \begin{bmatrix} -2 & 10 \\ 6 & -3 \\ -7 & -1 \end{bmatrix} \qquad C = \begin{bmatrix} 8 & -1 \\ 8 & -10 \\ -3 & -5 \end{bmatrix}$$

2. Given the following matrices, if possible, determine $3A$. If not, state "Not Possible".

$$A = \begin{bmatrix} -2 & 8 \\ -4 & -1 \\ 8 & 2 \end{bmatrix}$$

3. Given the following matrices, if possible, determine $3B - 3A$. If not, state "Not Possible".

$$A = \begin{bmatrix} 6 & 0 \\ -5 & 9 \\ 0 & -7 \end{bmatrix} \qquad B = \begin{bmatrix} -10 & -3 & 4 \\ -3 & 0 & -1 \end{bmatrix}$$

4. Determine values of the variables that will make the following equation true, if possible. If not, state "Not Possible".

$$2\begin{bmatrix} 3x & y & 2z \end{bmatrix} + 2\begin{bmatrix} -x & 3y & z \end{bmatrix} = \begin{bmatrix} 32 & -72 & 54 \end{bmatrix}$$

5. Determine values of the variables that will make the following equation true, if possible. If not, state "Not Possible".

$$2\begin{bmatrix} 3r \\ 3s \end{bmatrix} - 5\begin{bmatrix} -2s \\ r \end{bmatrix} = \begin{bmatrix} 78 \\ -22 \end{bmatrix}$$

6. Given the following matrices, if possible, determine AB. Identify the dimensions of the resulting matrix and fill out the matrix, if it exists. If not, state "Not Possible".

$$A = \begin{bmatrix} -5 & 0 & 0 \end{bmatrix} \qquad B = \begin{bmatrix} -3 & -7 \\ 3 & -8 \\ -3 & -7 \end{bmatrix}$$

7. Given the following matrices, if possible, determine AB. Identify the dimensions of the resulting matrix and fill out the matrix, if it exists. If not, state "Not Possible".

$$A = \begin{bmatrix} 8 \\ -4 \\ -9 \end{bmatrix} \qquad B = \begin{bmatrix} 9 & 8 & 2 \end{bmatrix}$$

8. Given the following matrices, if possible, determine AB. Identify the dimensions of the resulting matrix and fill out the matrix, if it exists. If not, state "Not Possible".

$$A = \begin{bmatrix} 4 & -10 \\ -2 & -5 \\ -6 & -9 \end{bmatrix} \qquad B = \begin{bmatrix} 3 & 7 & -10 \\ 6 & 0 & 8 \end{bmatrix}$$

9. Given the following matrices, if possible, determine $(BC)\,A$. Identify the dimensions of the resulting matrix and fill out the matrix, if it exists. If not, state "Not Possible".

$$A = \begin{bmatrix} 8 & -9 \\ -4 & -9 \end{bmatrix}, B = \begin{bmatrix} 3 & -3 \end{bmatrix}, \text{ and } C = \begin{bmatrix} -3 & -3 \\ -7 & 8 \end{bmatrix}$$

10. Determine values of the variables that will make the following equation true, if possible. If not, state "Not Possible".

$$3 \begin{bmatrix} 3r \\ 5 \end{bmatrix} + 2 \begin{bmatrix} -30 \\ 3s \end{bmatrix} = \begin{bmatrix} 3s \\ 3r \end{bmatrix}$$

6.5 Inverses of Square Matrices

EXAMPLE 1: Matrix Equation

Write each linear system as a matrix equation.

a. $\begin{cases} -3x + 5y = 2 \\ x - 4y = -1 \end{cases}$

b. $\begin{cases} 3y - x = -2 \\ 4 - z + y = 5 \\ z - 3x + 3 = y - x \end{cases}$

Solution

Identity Matrices

The $n \times n$ **identity matrix**, denoted I_n (just I when there is no possibility of confusion), is the $n \times n$ matrix consisting of 1's on the _____ and 0's _____.

The **main diagonal** consists of those entries in the _____, the _____, and so on down to the nth row- n^{th} column. Every identity matrix has the form

$$ I = \begin{bmatrix} 1 & 0 & 0 & \cdots & 0 \\ 0 & 1 & 0 & \cdots & 0 \\ 0 & 0 & 1 & \cdots & 0 \\ \vdots & \vdots & \vdots & \ddots & \vdots \\ 0 & 0 & 0 & \cdots & 1 \end{bmatrix}. $$

If the matrices A and B have appropriate order, so that the matrix products are defined, then $AI = A$ and $IB = B$. Thus, the identity matrix serves as the multiplicative _____ on the set of appropriately-sized matrices. In this sense, I serves the same purpose as the number 1 in the set of real numbers.

The Inverse of a Matrix

Let A be an $n \times n$ matrix. If there exists an $n \times n$ matrix A^{-1} such that

$$A^{-1}A = I_n \text{ and } AA^{-1} = I_n,$$

we call A^{-1} the _____ .

EXAMPLE 2: Finding the Inverse of a Matrix

Find the inverse of the matrix $A = \begin{bmatrix} 2 & -3 \\ -1 & 2 \end{bmatrix}$.

Solution

Finding the Inverse of a Matrix

Let A be an $n \times n$ matrix. The inverse of A can be found by:

Step 1:   Forming the augmented matrix $\begin{bmatrix} A & | & I \end{bmatrix}$, where I is the $n \times n$ identity matrix.

Step 2:   Using Gauss-Jordan elimination to put $\begin{bmatrix} A & | & I \end{bmatrix}$ into the form $\begin{bmatrix} I & | & B \end{bmatrix}$, if possible.

Step 3:   Defining A^{-1} to be B.

If it is not possible to put $\begin{bmatrix} A & | & I \end{bmatrix}$ into reduced row echelon form, then A doesn't have an inverse, and we say A is _____ .

If the coefficient matrix of a system of equations is not invertible, it means that the system either has an _____ number of solutions or has _____ .

Inverse of a 2 × 2 Matrix

Let $A = \begin{bmatrix} a & b \\ c & d \end{bmatrix}$. Then $A^{-1} = \dfrac{1}{|A|} \begin{bmatrix} d & -b \\ -c & a \end{bmatrix}$, where $|A| = ad - bc$ is the determinant of A. Since $|A|$ appears in the denominator of a fraction, A^{-1} fails to exist if $|A| = 0$.

Invertible Matrices

A square matrix A is _____ if and only if $|A| \neq 0$.

EXAMPLE 3: Finding the Inverse of a Matrix

Find the inverses of the following matrices, if possible.

a. $A = \begin{bmatrix} 3 & -5 \\ 2 & 1 \end{bmatrix}$

b. $B = \begin{bmatrix} 2 & 4 & 2 \\ -1 & 5 & -1 \\ 3 & 1 & 3 \end{bmatrix}$

Solution

The Inverse Matrix Method

To solve a linear system of n equations in n variables:

Step 1: Write the system in matrix form as $AX = B$, where A is the $n \times n$ matrix of _____ , X is the $n \times 1$ matrix of _____ , and B is the $n \times 1$ matrix of _____ .

Step 2: Calculate A^{-1}, if it exists. If A^{-1} does not exist, the system either has an _____ number of solutions, or _____ , and a different method is required.

Step 3: Multiply both sides of the equation $AX = B$ by A^{-1}. We obtain

$$A^{-1}AX = A^{-1}B$$
$$IX = A^{-1}B$$
$$X = A^{-1}B$$

Step 4: The entries in the $n \times 1$ matrix _____ are the solutions for the variables listed in the $n \times 1$ matrix X.

CAUTION!

It is crucial to multiply both sides of the equation $AX = B$ by A^{-1} on the _____ .
Recall that matrix multiplication is not commutative, so failing to do this can result in an incorrect answer, and often the multiplication will not even be defined.

EXAMPLE 4: Inverse Matrix Method

Solve the following systems by the inverse matrix method.

a. $\begin{cases} 4x - 5y = 3 \\ -3x + 7y = 1 \end{cases}$

b. $\begin{cases} -x + 2y = 3 \\ 3x - 6y = -5 \end{cases}$

Solution

6.5 Exercises

1. Write the following system of equations as a single matrix equation.

$$\begin{cases} 4x - 7y = 2 \\ 2x + 8y = -10 \end{cases}$$

2. Write the following system of equations as a single matrix equation.

$$\begin{cases} 5x - 3y + 2z = 5 \\ 9x + 8y = 0 \\ 9y - 4z = 9 \end{cases}$$

3. Find the inverse of the following matrix, if possible. Reduce all fractions to lowest terms.

$$A = \begin{bmatrix} -4 & -6 \\ -1 & 2 \end{bmatrix}$$

4. Find the inverse of the following matrix, if possible. Reduce all fractions to lowest terms.

$$A = \begin{bmatrix} -2 & 3 \\ 4 & -6 \end{bmatrix}$$

5. Find the inverse of the following matrix, if possible. Reduce all fractions to lowest terms.

$$A = \begin{bmatrix} -1 & 3 & 4 \\ 4 & -11 & -18 \\ 2 & -7 & -7 \end{bmatrix}$$

6. Solve the following system by the inverse matrix method, if possible. If the inverse matrix method doesn't apply, use any other method to determine if the system is inconsistent or dependent.

$$\begin{cases} -x - 4y = -7 \\ 5x + 2y = 17 \end{cases}$$

7. Solve the following system by the inverse matrix method, if possible. If the inverse matrix method doesn't apply, use any other method to determine if the system is inconsistent or dependent.

$$\begin{cases} 2x + 3y = -1 \\ 8x + 12y = -4 \end{cases}$$

8. Solve the following system by the inverse matrix method, if possible. If the inverse matrix method doesn't apply, use any other method to determine if the system is inconsistent or dependent.

$$\begin{cases} -2x - 4y = -6 \\ -6x - 12y = -21 \end{cases}$$

9. Solve the following system by the inverse matrix method, if possible. If the inverse matrix method doesn't apply, use any other method to determine if the system is inconsistent or dependent.

$$\begin{cases} 2y - 3z = -8 \\ -x - 3y + z = 10 \\ 3x - 7y - 4z = -16 \end{cases}$$

10. Solve the following system by the inverse matrix method, if possible. If the inverse matrix method doesn't apply, use any other method to determine if the system is inconsistent or dependent.

$$\begin{cases} 3x - 6y + 9z = -1 \\ 2x + 2y - 6z = -2 \\ x - 2y + 3z = -1 \end{cases}$$

6.6 Leontief Input-Output Analysis

Solving for a Production Matrix

Given a system of n industries, complete the following steps to solve for the production matrix X.

Step 1: Construct the $n \times 1$ _____ X.

$$X = \begin{bmatrix} x_1 \\ x_2 \\ \vdots \\ x_n \end{bmatrix}$$

- $x_i = $ the total production of the ith industry

- This step is basically a formality since X is not used in the following constructions and computations. It is, however, a good practice to write the matrix as it assists in keeping the production variables properly ordered.

Step 2: Construct the $n \times n$ _____ A.

$$A = \begin{bmatrix} a_{11} & a_{12} & \cdots & a_{1n} \\ a_{21} & a_{22} & \cdots & a_{2n} \\ \vdots & \vdots & \ddots & \vdots \\ a_{n1} & a_{n2} & \cdots & a_{nn} \end{bmatrix}$$

- $a_{ij} = $ the input needed from the ith industry for the jth industry to produce one unit of output

- The inputs are always between 0 and 1 since each input is the fraction of a unit needed from the ith industry to produce one unit of output for the jth industry.

Step 3: Construct the $n \times 1$ _____ E.

$$E = \begin{bmatrix} e_1 \\ e_2 \\ \vdots \\ e_n \end{bmatrix}$$

- $e_i = $ the external demand of the ith industry

Step 4: Calculate $X = (I - A)^{-1}E$ **where** I **is the** $n \times n$ _____

_____.

Note: If technology is available, this may be calculated in a very straightforward manner. Otherwise, completing the following steps may be necessary.

1. Calculate $I - A$.

2. Calculate $(I - A)^{-1}$.

3. Calculate $(I - A)^{-1}E$.

Step 5: Use the results of Step 4 to write the solution for the _____ _____ **and interpret the results in terms of units of production of each industry that are necessary to satisfy the** _____ _____ **demands of the system.**

Example 2: Closed Economy

Suppose a closed economy consists of three industries—with x_1 equal to the value of shipping output, x_2 equal to the value of manufacturing output, and x_3 equal to the government budget—and has the following input-output matrix.

$$A = \begin{matrix} & \begin{matrix} S & M & G \end{matrix} & \\ \begin{bmatrix} 0.4 & 0.2 & 0.2 \\ 0.2 & 0.3 & 0.3 \\ 0.4 & 0.5 & 0.5 \end{bmatrix} & \begin{matrix} \text{Shipping} \\ \text{Manufacturing} \\ \text{Government} \end{matrix} \end{matrix}$$

Find each industry's production (that is, find the shipping output x_1, manufacturing output x_2, and government budget x_3).

Solution

6.6 Exercises

1. Given the following input-output matrix A and corresponding demand matrix E for a local economy, find the production matrix X.

$$A = \begin{matrix} & \begin{matrix} \text{A} & \text{M} \end{matrix} \\ \begin{bmatrix} 0.7 & 0.1 \\ 0.2 & 0.6 \end{bmatrix} & \begin{matrix} \text{Agriculture} \\ \text{Manufacturing} \end{matrix} \end{matrix}, E = \begin{bmatrix} 8000 \\ 5000 \end{bmatrix}$$

2. Given the following input-output matrix A and corresponding demand matrix E for a local economy, find the production matrix X.

$$A = \begin{matrix} & \begin{matrix} \text{S} & \text{M} & \text{R} \end{matrix} \\ \begin{bmatrix} 0.2 & 0.3 & 0.3 \\ 0.1 & 0.2 & 0.2 \\ 0.1 & 0.1 & 0.1 \end{bmatrix} & \begin{matrix} \text{Single Family} \\ \text{Multi-Family} \\ \text{Rental} \end{matrix} \end{matrix}, E = \begin{bmatrix} 600 \\ 200 \\ 400 \end{bmatrix}$$

3. Suppose that in a certain local economy we have natural gas and coal industries. To produce one dollar in output, each industry needs the following input:

   o The natural gas industry requires $0.40 from itself and $0.30 from coal.

   o The coal industry requires $0.20 from natural gas and $0.40 from itself.

   Suppose further that in addition to the internal demand, there is a surplus demand from outside of the industries for $600 in natural gas and $900 in coal. Solve for the production necessary to meet these internal and surplus demands.

4. Suppose that in a certain local entertainment economy we have museums, theaters, and sporting events industries. To produce one dollar in output, each industry needs the following input:

o The museum industry requires $0.40 from itself, $0.30 from theater, and $0.20 from sporting events.

o The theater industry requires $0.20 from museums, $0.40 from itself, and $0.30 from sporting events.

o The sporting events industry requires $0.20 from museums, $0.40 from theater, and $0.30 from itself.

Suppose further that in addition to the internal demand, there is a surplus demand from outside of the industries for $500 in museums, $750 in theaters, and $400 in sporting events. Solve for the production necessary to meet these internal and surplus demands.

5. Given the following input-output matrix A and corresponding production matrix X, find the external demand matrix E.

$$A = \begin{bmatrix} \overset{C}{0.2} & \overset{Au}{0.3} & \overset{Ai}{0.1} \\ 0.1 & 0.4 & 0.2 \\ 0.6 & 0.2 & 0.3 \end{bmatrix} \begin{matrix} \text{Computers} \\ \text{Automobiles} \\ \text{Aircraft} \end{matrix}, X = \begin{bmatrix} 800 \\ 600 \\ 1000 \end{bmatrix}$$

6. Suppose that in a certain local economy we have durable manufacturing and nondurable manufacturing industries. To produce one dollar in output, each industry needs the following input:

o The durable manufacturing industry requires $0.10 from itself and $0.30 from nondurable manufacturing.

o The nondurable manufacturing industry requires $0.30 from durable manufacturing and $0.60 from itself.

Suppose further that total production capacity of durable manufacturing is $500 and of nondurable manufacturing is $800. Find the external demand.

7. Given the input-output matrix, determine which industry is most dependent on the NATURAL GAS industry.

$$A = \begin{bmatrix} \overset{C}{0.3} & \overset{G}{0.2} & \overset{E}{0.1} & \overset{N}{0.2} \\ 0.2 & 0.1 & 0.1 & 0.1 \\ 0.1 & 0.1 & 0.2 & 0.1 \\ 0.2 & 0.1 & 0.4 & 0 \end{bmatrix} \begin{matrix} \text{Coal} \\ \text{Gasoline} \\ \text{Electric} \\ \text{Natural Gas} \end{matrix}$$

CHAPTER 7

Inequalities and Linear Programming

7.1 Linear Inequalities in Two Variables

Solving Linear Inequalities in Two Variables

Step 1: Graph the _____ that results from replacing the inequality symbol with $=$.

Step 2: Make the line _____ if the inequality symbol is $\leq$ or $\geq$ (non-strict) and

_____ if the symbol is $<$ or $>$ (strict). A solid line indicates that points on the

line are _____ in the solution set while a dashed line indicates that points

on the line are _____ from the solution set.

Step 3: Determine which of the half-planes defined by the boundary line solves the inequality

by substituting a _____ from one of the two half-planes into the

inequality. If the resulting numerical statement is _____ , all the points in the same

half-plane as the test point solve the inequality. Otherwise, the points in the other half-

plane solve the inequality. Shade in the half-plane that _____ the inequality.

EXAMPLE 1: Solving Linear Inequalities

Solve the following linear inequalities by graphing their solution sets.

a. $3x + 2y < 12$

b. $x - y \leq 0$

c. $x > 3$

> **Note:**
>
> Whenever possible, use the origin as a test point, since it makes for a very simple calculation.

Solution

EXAMPLE 2: Solving Linear Inequalities

Graph the solution sets that satisfy the following inequalities.

a. $5x - 2y < 10$ and $y \leq x$

b. $x + y < 4$ or $x \geq 4$

Solution

EXAMPLE 3: Solving Absolute Value Linear Inequalities

Graph the solution set in $\mathbb{R}^2$ that satisfies the joint conditions $|x - 3| > 1$ and $|y - 2| \leq 3$.

Solution

EXAMPLE 4: Graphing Regions of Constraint

Graph the region of constraint defined by the following inequalities:

$$x \le 2, \, x + y \ge -3 \text{ and } y \le \frac{1}{2}x + 1$$

Note:

Linear programming problems usually involve inequalities which are not strict.

Solution

7.1 Exercises

1. Graph the solution set of the following linear inequality:

$$2x + 2y < -6$$

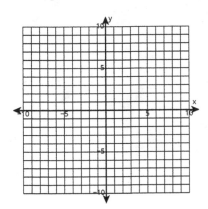

2. Graph the solution set of the following linear inequality:

$3y - 3x > 18$

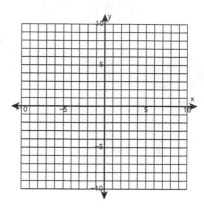

3. Graph the solution set of the following linear inequality:

$5y - 35 \leq -7x$

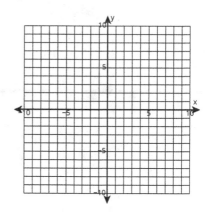

4. Graph the solution set of the following linear inequality:

$8y + 7x \geq 8y - 56$

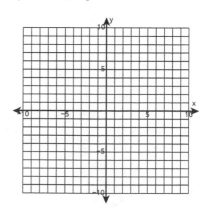

5. Solve the system of two linear inequalities graphically.

$5x + 6y < -30$ and $x \geq 2$

    a. Graph the solution set of the **first** linear inequality.

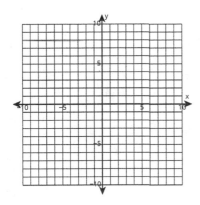

    b. Graph the solution set of the **second** linear inequality.

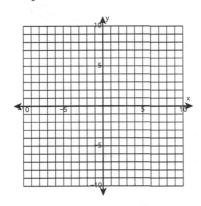

    c. Graph the solution set for the system.

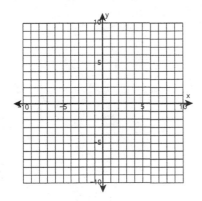

6. Solve the system of two linear inequalities graphically.

$4y - 4x > 24$ or $y > 2$

    a. Graph the solution set of the **first** linear inequality.

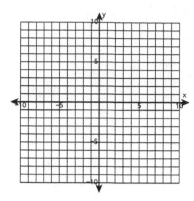

    b. Graph the solution set of the **second** linear inequality.

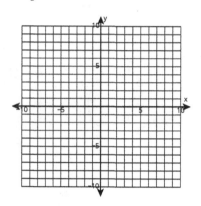

    c. Graph the solution set for the system.

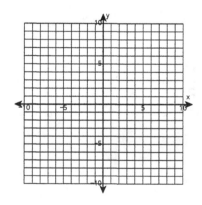

7. Consider the following absolute value inequality.

$|7x + 1| < 9$

    a. Rewrite the given inequality as two linear inequalities.

    b. Graph the solution set of theoriginal inequality.

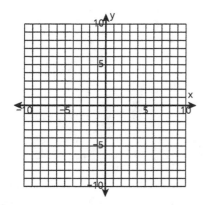

8. Consider the following absolute value inequality.

$|5y + 5| \geq 24$

    a. Rewrite the given inequality as two linear inequalities.

    b. Graph the solution set of the original inequality.

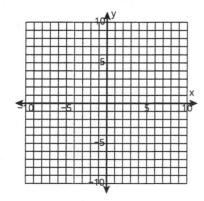

9. Solve the system of two linear inequalities graphically.

$$\begin{cases} 3y < -4x + 18 \\ 2y \geq 5x - 15 \end{cases}$$

a. Graph the solution set of the **first** linear inequality.

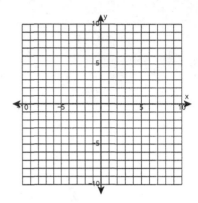

b. Graph the solution set of the **second** linear inequality.

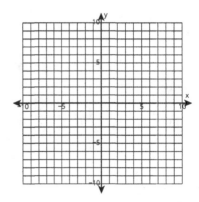

c. Graph the solution set for the system.

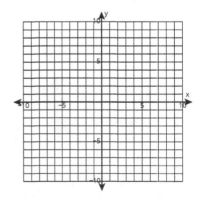

10. Solve the system of two linear inequalities graphically.

$$\begin{cases} y \leq -5x + 10 \\ \quad y > x - 2 \end{cases}$$

    a. Graph the solution set of the **first** linear inequality.

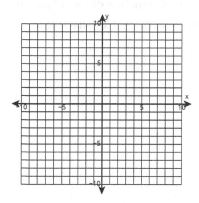

    b. Graph the solution set of the **second** linear inequality.

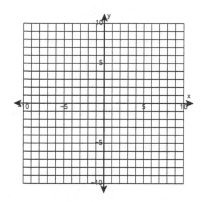

    c. Graph the solution set for the system.

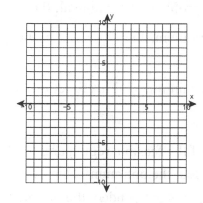

7.2 Linear Programming: The Graphical Approach

EXAMPLE 1: Feasible Regions

A family orchard sells peaches and nectarines. To prevent a pest infestation, the number of nectarine trees in the orchard cannot exceed the number of peach trees. Also, because of the space requirements of each type of tree, the number of nectarine trees plus twice the number of peach trees cannot exceed 100 trees. Construct the constraints and graph the feasible region for this situation.

Solution

Bounded and Unbounded Regions

A _____ region is one in which all the points lie within some finite distance of the origin.

An _____ region has points that are arbitrarily far away from the origin.

Linear Programming

Step 1: Identify the _____ to be considered in the problem, determine all the _____ on the variables imposed by the problem, and _____ the feasible region described by the constraints.

Step 2: Determine the function that is to be either maximized or minimized. This function is called the _____ .

Step 3: Evaluate the function at each of the _____ of the feasible region and compare the values found. If the feasible region is bounded, the optimum value of the function will occur at a _____ . If the feasible region is unbounded, the optimum value of the function _____ , but if it does exist, it will occur at a _____ .

EXAMPLE 2: Linear Programming

Find the maximum and minimum values of $f(x, y) = 3x + 2y$ subject to the constraints

$$\begin{cases} x \geq 0 \\ y \geq 0 \\ y \geq 3 - 3x \\ 2x + y \leq 10 \\ x + y \leq 7 \\ y \geq x - 2 \end{cases}$$

Solution

EXAMPLE 3: Linear Programming

Find the maximum and minimum values of $f(x, y) = x + 2y$ subject to the constraints

$$\begin{cases} x \geq 0 \\ y \geq 3 - x \\ x - y \leq 1 \\ y \leq 2x + 4 \end{cases}$$

Solution

EXAMPLE 4: Linear Programming

A manufacturer makes two models of MP3 players. Model A requires 3 minutes to assemble, 4 minutes to test, and 1 minute to package. Model B requires 4 minutes to assemble, 3 minutes to test, and 6 minutes to package. The manufacturer can allot 7400 minutes total for assembly, 8000 minutes for testing, and 9000 minutes for packaging. Model A generates a profit of $7.00 and model B generates a profit of $8.00. How many of each model should be made in order to maximize profit?

Solution

7.2 Exercises

1. Find the minimum and maximum values of the objective function, and the points at which these values occur subject to the given constraints.

<u>Objective Function</u>   <u>Constraints</u>

$f(x, y) = 8x + 4y$ $x \geq 0$

                           $y \geq 0$

                     $x + y \leq 2$

2. Find the minimum and maximum values of the objective function, and the points at which these values occur subject to the given constraints.

<u>Objective Function</u>   <u>Constraints</u>

$f(x, y) = 5x + 5y$ $x \geq 0$

                           $y \geq 0$

                     $3x + y \leq 12$

3. Find the minimum and maximum values of the objective function, and the points at which these values occur subject to the given constraints. Round your answers to the nearest hundredth.

<u>Objective Function</u>   <u>Constraints</u>

$f(x, y) = 2x + 9y$ $0 \leq x \leq 2$

                           $0 \leq y \leq 4$

                     $9x + 8y \leq 36$

4. Find the minimum and maximum values of the objective function, and the points at which these values occur subject to the given constraints. Round your answers to the nearest hundredth.

Objective Function	Constraints
$f(x, y) = 2x + 9y$	$x \geq 0$
	$9x + 18y \geq 162$
	$2x - y \leq 10$
	$x + 6y \leq 84$

5. Find the minimum and maximum values of the objective function, and the points at which these values occur subject to the given constraints.

Objective Function	Constraints
$f(x, y) = 6x + 9y$	$x \geq 0$
	$y \geq 0$
	$x + y \leq 8$

6. A volunteer has been asked to drop off some supplies at a facility housing victims of a hurricane evacuation. The volunteer would like to bring at least 90 bottles of water, 56 first aid kits, and 40 security blankets on his visit. The relief organization has a standing agreement with two companies that provide victim packages. Company A can provide packages of 10 water bottles, 6 first aid kits, and 4 security blankets at a cost of $2.50. Company B can provide packages of 9 water bottles, 4 first aid kits, and 3 security blankets at a cost of $1.50. How many of each package should the volunteer pick up to minimize the cost? What amount does the relief organization pay?

7. To prevent pests, an orchard can have no more than 6 times as many apple trees as peach trees. Also, the number of apple trees plus 3 times the number of peach trees must not exceed 405. The revenue from a single apple tree is $110 and the revenue from a single peach tree is $117. Determine the number of each type of tree that will maximize revenue. What is the maximum revenue? Let x represent the number of apple trees and y represent the number of peach trees.

8. On your birthday your rich uncle gave you $13000. You would like to invest at least $4000 of the money in municipal bonds yielding 4% and no more than $3000 in Treasury bills yielding 7%. How much should be placed in each investment in order to maximize the interest earned in one year? Assume simple interest applies. Let x represent the amount of money in municipal bonds and y represent the amount of money in Treasury bills.

9. Find the minimum and maximum values of the objective function, and the points at which these values occur subject to the given constraints. Be sure to indicate in your answer if the minimum or maximum do not exist. Round your answers to the nearest hundredth.

Objective Function	Constraints
$f(x, y) = 2x + y$	$x \geq 0$
	$y \geq 7 - x$
	$x - y \leq 5$
	$y \leq 2x + 10$

10. A plane carrying food and water to a resort island can carry a maximum of 30000 pounds and is limited in space to carrying no more than 575 cubic feet. Each container of water weighs 60 pounds and takes up 1 cubic foot of cargo space. Each container of food weighs 175 pounds and takes up 5 cubic feet. Hotels on the island will buy the food for 17 dollars a pound and the water for 5 dollars per pound. What is the optimum number of containers of each item that will maximize the revenue generated by the plane? What is the maximum revenue? Let x represent the number of food containers and y represent the number of water containers.

7.3 The Simplex Method: Maximization

The Simplex Method

1. Rewrite the _____ as an equation in standard form with all variables on the left side and a nonnegative constant on the right side.

2. Convert each _____ into an equation with all variables on the left side and a positive constant on the right side by introducing a different slack variable for each constraint.

3. Form the system to solve with the constraint equations in the _____ and the objective function in the _____ .

4. Create the initial _____ by forming the augmented matrix for the system of equations with the objective function in the bottom row.

5. Locate the negative entry in the bottom row with _____ to determine the pivot column.

6. In each row except for the bottom one, divide the entry in the far-right column by the one in the pivot column. The row with the _____ will be the pivot row.

7. Perform elementary row operations on the simplex tableau to make the pivot element 1 and the remaining entries in the pivot column 0. Always use _____ to do any row operations.

8. If all entries in the bottom row are _____ , stop. If not, repeat steps 5 through 7 until all entries in the bottom row are nonnegative.

9. The _____ of the objective function is in the lower-right corner of the final simplex tableau.

Example 2: An Application of the Simplex Method

An amusement park vendor sells two types of bagged snack mix: Chocolaty Crunch and Pretzel Delight. Each bag of Chocolaty Crunch consists of 2 cups of M&M's, 2 cups of chocolate chips, and 1 cup of pretzels; and each bag of Pretzel Delight consists of 1 cup of M&M's, 2 cups of chocolate chips, and 2 cups of pretzels. Suppose the vendor has 70 cups of M&M's, 90 cups of chocolate chips, and 65 cups of pretzels in his inventory. Suppose also that one bag of Chocolaty Crunch sells for $15 and one bag of Pretzel Delight sells for $10. How many bags of each mix should the vendor sell in order to maximize revenue.

Solution

Example 4: A Maximization Problem with Multiple Solutions

Use the simplex method to maximize $f(x_1, x_2) = 6x_1 + 8x_2$ subject to the following constraints.

$$\begin{cases} 3x_1 + 4x_2 \le 48 \\ 2x_1 + 5x_2 \le 80 \end{cases}$$

Solution

Example 5: A Maximization Problem with No Solution

Use the simplex method to maximize $f(x_1, x_2, x_3) = x_1 + x_2 + 2x_3$ subject to the following constraints.

$$\begin{cases} x_1 - 2x_2 + x_3 \leq 2 \\ x_1 - 4x_2 + 2x_3 \leq 4 \end{cases}$$

Solution

7.3 Exercises

1. Maximize $f(x_1, x_2) = 5x_1 + 6x_2$ subject to the following constraints.

$$\begin{cases} x_1 + 2x_2 \leq 10 \\ 3x_1 + x_2 \leq 15 \end{cases}$$

2. Maximize $f(x_1, x_2) = 3x_1 + x_2$ subject to the following constraints.

$$\begin{cases} x_1 + 4x_2 \leq 5 \\ 2x_1 + x_2 \leq 3 \\ 3x_1 + 2x_2 \leq 5 \end{cases}$$

3. Maximize $f(x_1, x_2) = 4x_1 + 8x_2$ subject to the following constraints.

$$\begin{cases} 3x_1 + x_2 \leq 6 \\ x_1 + 2x_2 \leq 4 \end{cases}$$

4. Maximize $f(x_1, x_2, x_3) = 5x_1 + 7x_2 + 2x_3$ subject to the following constraints.

$$\begin{cases} x_1 + 2x_2 - x_3 \leq 5 \\ 2x_1 + x_2 - 2x_3 \leq 4 \end{cases}$$

5. Maximize $f(x_1, x_2, x_3) = 5x_1 + 6x_2 + 3x_3$ subject to the following constraints.

$$\begin{cases} 3x_1 + x_2 + x_3 \leq 18 \\ x_1 + 4x_2 + x_3 \leq 18 \\ x_1 + x_2 + 2x_3 \leq 19 \end{cases}$$

7.4 The Simplex Method: Duality and Minimization

Example 1

Find the minimum value of $f(x_1, x_2) = 5x_1 + 4x_2$ subject to the following constraints.

$$\begin{cases} 10x_1 + 2x_2 \geq 15 \\ x_1 + 2x_2 \geq 6 \end{cases}$$

Solution

Theorem of Duality

If a minimization problem and its dual maximization problem have a solution, the minimum value of the _____ is the same as the maximum value of the dual problem. Moreover, when the simplex method is used to solve the dual maximization problem, the coordinates at which the minimum value is attained is given by the bottom entry of the columns corresponding to the _____ _____ of the maximization problem.

Solving Minimization Problems

1. Form the matrix for minimization with constraints, A, for the _____ .

2. Find _____ , and relabel the objective function in the lower-right.

3. Write down the _____ maximization problem. Use variable names different from those in the primal problem.

4. Create the _____ by forming the augmented matrix for the system of equations with the objective function in the bottom row.

5. Solve the dual maximization problem using the _____ . Use the same procedure as in the previous lesson.

6. The _____ of the objective function is in the lower-right corner of the final simplex tableau, and the bottom entry in the columns corresponding to the _____ _____ give the coordinates at which the minimum is attained.

Example 2: Minimizing Cost

A tool manufacturing company produces drills and saws at two different factories. At factory A, 100 drills and 150 saws can be manufactured in one hour, and the hourly cost is $800. In each hour at factory B, 200 drills and 150 saws are manufactured in one hour, and the hourly cost is $1000. The company is expecting orders each week for at least 8000 drills and 10,000 saws. How many hours per week should each factory be operated in order to provide the inventory for these orders at the minimum cost?

Solution

Example 3: Low-Carb Diet

Suppose a salad bar has three different options of salads. The table below gives the protein content (in grams), vitamin C content (in milligrams), calorie count, and carbohydrate content (in grams) in a 2.5-ounce serving of each salad.

Salad	Protein	Vitamin C	Calorie Count	Carbs
A	6 g	40 mg	60	16 g
B	4 g	10 mg	30	8 g
C	2 g	30 mg	120	12 g

Suppose that you are on a low-carb diet in which you need a minimum of 12 grams of protein, 70 milligrams of vitamin C, and 240 calories. How many 2.5 ounce servings of each salad should you eat at this salad bar to minimize the carbohydrate content?

Solution

Simplex Method: Maximization (7.3)	Duality Principle: Minimization (7.4)
1.) Maximize $f(x_1, x_2, \ldots, x_n)$ subject to	1.) Minimize $f(x_1, x_2, \ldots, x_n)$ subject to
$a_1 x_1 + a_2 x_2 + \ldots + a_n x_n \leq c.$	_____ .
	2.) Write down the matrix for minimization
	with constraints, A.
	3.) Write down the matrix for the dual
	maximization problem, _____ .
2.) Introduce slack variables $s_1, s_2, \ldots, s_m$,	4.) Introduce slack variables $s_1, s_2, \ldots, s_m$
and form initial simplex tableau.	and form _____ .
3.) Repeat pivoting process until optimal	5.) Repeat pivoting process until optimal
solution is reached.	solution to dual problem is reached.
4.) Look at the final simplex tableau. The	6.) Look at the final simplex tableau. The
maximum value of f is in lower-right	_____ of f is in lower-right
corner. The nonzero coordinates at which	corner. The coordinates at which the
the maximum is attained are in the last	minimum is attained are in the last row
column in the rows corresponding to the	of the columns corresponding to the
1 entry of each x_i that is basic, and	_____ .
each x_i that is nonbasic is 0.	

7.4 Exercises

1. Minimize $f(x_1 x_2) = 3x_1 + x_2$ subject to the following constraints.

$$\begin{cases} 4x_1 + 5x_2 \geq 13 \\ 2x_1 + x_2 \geq 5 \end{cases}$$

2. Minimize $f(x_1, x_2) = 2x_1 + 4x_2$ subject to the following constraints.

$$\begin{cases} 3x_1 + x_2 \geq 6 \\ x_1 + x_2 \geq 4 \end{cases}$$

3. Minimize $f(x_1, x_2, x_3) = x_1 + 2x_2 + x_3$ subject to the following constraints.

$$\begin{cases} x_1 + 3x_2 \geq 6 \\ 2x_1 + x_3 \geq 10 \end{cases}$$

4. Minimize $f(x_1, x_2, x_3) = 11x_1 + 8x_2 + 7x_3$ subject to the following constraints.

$$\begin{cases} x_1 + 2x_2 + 2x_3 \geq 8 \\ x_1 + 2x_2 + x_3 \geq 6 \\ 4x_1 + x_2 + x_3 \geq 4 \end{cases}$$

5. Minimize $f(x_1, x_2, x_3) = 7x_1 + 10x_3 + 7x_3$ subject to the following constraints.

$$\begin{cases} x_1 + x_2 + 2x_3 \geq 5 \\ x_1 + 3x_2 \geq 6 \\ 2x_1 + x_2 + x_3 \geq 4 \end{cases}$$

7.5 The Simplex Method: Mixed Constraints

Example 1: A Maximization Problem with Mixed Constraints

Maximize the objective function $f(x_1, x_2) = 5x_1 + 7x_2$ subject to the following constraints.

$$\begin{cases} 8x_1 + 4x_2 \leq 28 \\ -3x_1 + 2x_2 \geq 7 \end{cases}$$

Solution

Example 2: A Minimization Problem with Mixed Constraints

Minimize $f(x_1, x_2) = 4x_1 + 3x_2$ subject to the following constraints.

$$\begin{cases} 6x_1 - 5x_2 \leq 4 \\ 2x_1 + 5x_2 \geq 8 \end{cases}$$

Solution

Solving Mixed Constraint Problems Using the Simplex Method

1. If the optimization problem involves minimizing an objective function, f, view it as _____ $-f$.

2. For each constraint with the $\geq$ inequality (if any), multiply by _____ to change it to a $\leq$ inequality.

3. Introduce slack variables and create the preliminary simplex tableau by forming the _____ _____ for the system of equations with the objective function in the bottom row.

4. If there are any negative entries in the last column above the bottom row, use the following steps.

    a. Choose any other negative value in the same row and use its column as the

    _____ .

    b. In each row except for the _____ , divide the entry in the far-right column by the one in the pivot column. The row with the _____ will be the pivot row.

    c. Complete the iteration of the pivoting process, and repeat (a) and (b) until there are no negative entries in the _____ .

5. Apply the _____ to the resulting initial tableau, and arrive at an optimal solution.

Solving Mixed Constraint Problems Using the Big M Method

1. If the optimization problem involves minimizing an objective function, f, view it as maximizing _____ .

2. Introduce slack variables $s_i \geq 0$ in each constraint involving the _____ inequality.

3. Introduce surplus variables $s_j \geq 0$ (where $j > i$) in each constraint involving the _____ inequality.

4. Introduce artificial variables $a_k \geq 0$ in each constraint which originally involved a _____ or _____ (before step 3 was done).

5. For each artificial variable a_k, subtract a term of Ma_k from the function to be maximized, where $M > 0$ is a _____ . Use the same coefficient, M, in each of these terms.

6. Set z equal to the function to be maximized and rewrite it in form where the left side of the equation includes all variables and the right side is _____ .

7. Form the preliminary _____ with the constraints in the top rows and the objective function in the bottom row.

8. Perform any row operations to eliminate _____ at the bottom entry of each column corresponding to an artificial variable, and obtain a second simplex tableau.

9. Apply the _____ to the resulting initial tableau, and arrive at an optimal solution.

Example 4: An Application Using the Big M Method to Minimize Cost

Suppose a company produces silverware at three different plants. Plant A can manufacture 1500 forks, 2000 spoons, and 2500 knives in an hour, and the hourly cost of this plant is $600 per hour. Plant B does not manufacture forks, but it can manufacture 1800 spoons and 3000 knives in an hour. The hourly cost for plant B is $400 per hour. Plant C does not manufacture knives, but it can manufacture 2400 forks and 1800 spoons in an hour. The hourly cost for plant C is $500 per hour. The company is preparing an order this upcoming week for a large event at which food is being served to thousands of people. The organizers of this event are requesting at least 15000 forks, exactly 15000 spoons, and no more than 9000 knives. How many hours this coming week should each plant be operated in order to fulfill the demands of the event organizers at a minimum cost to the company?

Solution

7.5 Exercises

1. Minimize $f(x_1, x_2) = 4x_1 + 6x_2$ subject to the following constraints.

$$\begin{cases} 2x_1 + 3x_2 \leq 10 \\ 4x_1 - x_2 \geq 4 \end{cases}$$

2. Maximize $f(x_1, x_2, x_3) = 3x_1 + 2x_2 + 4x_3$ subject to the following constraints.

$$\begin{cases} 2x_1 + x_2 + x_3 \leq 7 \\ 3x_1 - x_2 + 4x_3 \geq 4 \\ x_1 + x_3 \geq 5 \end{cases}$$

3. Maximize $f(x_1, x_2, x_3) = 2x_1 - 2x_2 + 6x_3$ subject to the following constraints.

$$\begin{cases} x_1 + x_2 \leq 20 \\ x_1 + x_3 = 5 \\ x_2 + x_3 \geq 10 \end{cases}$$

4. Minimize $f(x_1, x_2) = 3x_1 + 4x_2$ subject to the following constraints.

$$\begin{cases} 4x_1 + 5x_2 \geq 18 \\ x_1 + x_2 = 4 \end{cases}$$

5. Minimize $f(x_1, x_2, x_3) = 7x_1 + 2x_2 + 3x_3$ subject to the following constraints.

$$\begin{cases} x_1 + x_2 + x_3 \geq 6 \\ 3x_1 - x_2 + 2x_3 = 7 \\ -x_1 + x_3 \leq 2 \end{cases}$$

CHAPTER 8

Probability

8.1 Set Notation

Set

A _____ is a collection of objects made up of specified elements, or members.

Roster Method

_____ is a way to describe a set by listing all of the elements in the set.

Equal Sets

Two sets are said to be _____ if they contain exactly the same elements. If sets A and B are equal, we write $A = B$.

Cardinal Number

The number of elements contained in a finite set is called the _____, or _____. The cardinal number is denoted by $|\ |$.

Equivalent Sets

Sets are _____ if they have the same cardinal number; that is, the same number of elements. If sets C and D are equivalent, we write $C \sim D$.

Example 2: Determining Equal and Equivalent Sets

Determine if the given pairs of sets are equal, equivalent, or neither.

a. $A = \{\text{Public Health, International Studies, Mechanical Engineering, Music Education, Political Science}\}$

$B = \{\text{Tim, Gloria, Alan, Warren, Karen}\}$

b. $X = \{2, 4, 6, 8, 10, 12, 14, 16, 18, 20\}$

$Y = \{20, 18, 16, 14, 12, 10, 8, 6, 4, 2\}$

Solution

Set-Builder Notation

_____ is used to describe a set when the members all share certain properties.

Empty Set

The _____ , or **null set**, is the set that contains no elements. If set A is empty, we write $A = \varnothing$.

Example 4: Determining Empty Sets

Determine if the following sets are empty sets.

a. The set A of negative numbers less than 100.

b. The set B of any state that contains the letter q in its name.

Solution

Universal Set

The set of all elements being considered for any particular situation is called the _____ _____ and is denoted by U.

Complement

The _____ of A consists of all the elements in the given universal set that are not contained in A. The complement of A is denoted A'.

8.1 Exercises

1. Determine the cardinal number of the set.

$W = \{t | t \in \mathbb{N}, t \text{ is even, and } |t| < 9\}$

2. Determine the cardinal number of the set.

$W = \{m, x, q, f, k, b, s, w, r, e, j, c\}$

3. Use the given sets to answer the question. Is $P = Q$? Why or why not?

$P = \{\text{September, April, June, November, February}\}$

$Q = \{\text{September, August, November, June, July}\}$

4. Write the set using the roster method.

Set C is the set of two-digit even numbers greater than 72 that do not contain the digit 8.

5. Use the variable x to write the set in set-builder notation.

Let G be the set of real numbers greater than ten.

6. Identify the term that is defined in the given statement.

A set that contains no elements.

7. Identify the term that is defined in the given statement.

A and B are two sets that have the same number of elements.

8. Let *C* equal the set of countries who have won more than 200 Silver medals. Write the set *C* using the roster method. Let the universal set consist of the 10 countries listed in the table. Use the country abbreviations US, SU, GB, G, F, I, S, C, R, and EG when writing the set.

Top 10 All-Time Olympic Medal Winning Countries				
Team	**Gold**	**Silver**	**Bronze**	**Combined Total**
United States (US)	1072	860	749	2681
Soviet Union (SU)	473	376	355	1204
Great Britain (GB)	246	276	284	806
Germany (G)	252	260	270	782
France (F)	233	254	293	780
Italy (I)	235	200	228	663
Sweden (S)	193	204	230	627
China (C)	213	166	147	526
Russia (R)	182	162	177	521
East Germany (EG)	192	165	162	519

Source: Wikipedia, s.v. "All-time Olympic Games medal table," accessed July 2014,http://en.wikipedia.org/wiki/All-time_Olympic_Games_medal_table

9. If $A = \{v, a, n, q, u, i, s, h, e, d\}$ and $U = \{a, b, c, d, e, f, g, h, i, j, k, l, m, n, o, p, q, r, s, t, u, v, w, x, y, z\}$, find A'.

10. If $A = \{i, n, s, p, e, c, t, o, r\}$ and $U = \{a, b, c, d, e, f, g, h, i, j, k, l, m, n, o, p, q, r, s, t, u, v, w, x, y, z\}$, find $|A'|$.

8.2 Operations with Sets

Intersection

The _____ of two sets A and B is the set of all elements common to both A and B. We denote the intersection of A and B as $A \cap B = \{x | x \in A \text{ and } x \in B\}$.

Example 2: Using a Venn Diagram to Find the Intersection

The given Venn diagram represents the number of students that participated in certain activities while on a spring break trip. Determine the number of students that went both hiking and skiing over spring break.

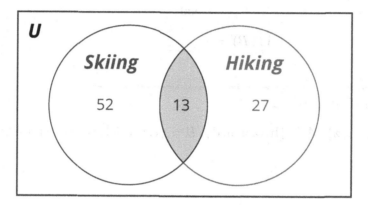

Solution

Union

The _____ of two sets A and B is the set of all elements in A or in B. We denote the union of A and B as $A \cup B = \{x | x \in A \text{ or } x \in B\}$.

Example 4: Combining Intersection and Union

Let $U = \{a, b, c, d, e, \ldots, z\}$, $M = \{m, a, t, h\}$, $N = \{m, o, n, e, y\}$, and $K = \{i, n, v, e, s, t, o, r\}$. Find

a. $M \cup (N \cap K)$

b. $M \cap (N \cup K)$

Solution

Disjoint

Two sets A and B are _____ if there are no elements in set A that are also contained in set B.

In the case where two sets are disjoint, the intersection of those two sets is the **null**, or empty set $\varnothing$. Therefore, $A \cap B = \varnothing$ when sets A and B are disjoint.

De Morgan's Laws

Let A and B be sets. Then,

$$(A \cup B)' = \underline{\hspace{3cm}}$$

and

$$(A \cap B)' = \underline{\hspace{3cm}}$$

Example 6: Using De Morgan's Laws

Given sets $U = \{a, b, c, d, \ldots, z\}$, $A = \{h, o, u, n, d\}$, $B = \{r, o, c, k\}$, verify that $(A \cup B)' = A' \cap B'$.

Solution

Example 7: Determining the Cardinal Number of a Union

Let $A = \{1, 2, 3, 4, 5\}$ and $B = \{2, 4, 6, 8\}$. Find $|A \cup B|$.

Solution

Inclusion-Exclusion Principle

The _____ states that the number of elements in the union of two sets A and B is calculated by adding the number of elements in set A to the number of elements in set B, less the number of elements that appear in both sets. We denote this by $|A \cup B| = |A| + |B| - |A \cap B|$.

8.2 Exercises

1. Use the given sets to find $A \cup B$.

$A = \{1, 2, 3, 4, 5, 6, 7, 8\}$

$B = \{4, 6, 8, 10, 12, 14\}$

2. Use the given sets to find $A \cap B$.

$A = \{4, 9, 14, 19\}$

$B = \{1, 2, 3, 15, 16, 17, 18\}$

3. Use the given sets to find $P \cup (Q \cap R)$.

$U = \{a, b, c, d, \ldots, x, y, z\}$

$P = \{s, a, m, p, l, e\}$

$Q = \{f, r, u, i, t\}$

$R = \{e, x, c, l, a, i, m\}$

4. Use the given sets to find $|P \cap Q|$.

$U = \{a, b, c, d, \ldots, x, y, z\}$

$P = \{b, l, u, e\}$

$Q = \{p, u, r, e, l, y\}$

5. In the following Venn diagram, *U* is the set of students in a class, *A* is the set of students who are majoring in psychology, and *B* is the set of students who are majoring in business. Determine which students are majoring in only psychology.

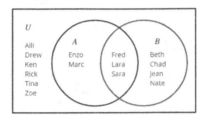

6. In the following Venn diagram, *U* is the set of students in a class, *A* is the set of students who have a tablet, and *B* is the set of students who have a graphing calculator. Determine which students have a tablet and a graphing calculator.

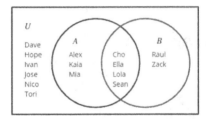

7. In the following Venn diagram, *U* is the set of students in a class, *A* is the set of students who have a tablet, and *B* is the set of students who have a graphing calculator. Determine which students have a tablet or a graphing calculator.

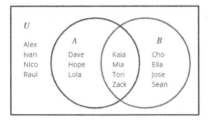

8. Find the number of playing cards in a standard deck of 52 cards that are either diamonds or cards that are not face cards.

9. Identify the term that best matches the given definition.

The set that contains all elements common to both sets A and B and denoted as
$A \cap B = \{x \mid x \in A \text{ and } x \in B\}$

10. In the following Venn diagram, U is the set of students in a class, A is the set of students who play soccer, and B is the set of students who play baseball. Determine how many students play soccer or baseball.

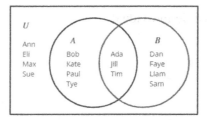

8.3 Introduction to Probability

Trial

A **trial**, or _____ , is any process that produces a random result.

Outcome

The **outcomes** of a trial are the possible individual _____ .

Sample Space

The **sample space** S of a trial is the set of all _____ outcomes.

Event

An **event** E is a group, or subset, of _____ the sample space.

Classical Probability

If all outcomes are _____ , **classical probability** is calculated with the formula

$$P(\text{event}) = \frac{\text{the number of possible outcomes in the event}}{\text{the number of outcomes in the sample space}}.$$

$P(\text{event})$ will always be a real number between 0 and 1, inclusive.

Example 3: Calculating Classical Probability

Suppose that you grab a snack from a bag of chocolates that contains 4 caramel with milk chocolate, 4 peppermint with white chocolate, 6 dark chocolate with mint, and 2 raspberry with dark chocolate. What is the probability that you randomly grab a raspberry with dark chocolate for your snack?

Solution

Empirical Probability

If all outcomes are based on an _____, **empirical probability** is calculated with the formula

$$P(\text{event}) = \frac{\text{the number of times the event occurs}}{\text{total number of times the experiment is performed}}.$$

$P(\text{event})$ will always be a real number between 0 and 1, inclusive.

Example 5: Classical vs. Empirical Probability

Determine if the scenarios given are examples of classical or empirical probability techniques.

a. Katie is curious about her chances of winning an e-reader from the student government association. She polled her friends to find out how many of them filled out the survey to be entered in the contest.

b. Tristan is interested in his chances of winning at the black jack table. He determines the probability of what his next card will be by knowing the cards that have already been played.

c. Based on the recent United States Census, the local government estimates the amount of growth the community will experience in the coming years.

Solution

8.3 Exercises

1. When planning a cruise, you have a choice of 2 destinations: Cozumel (C) or Jamaica (J); a choice of 4 types of rooms: balcony (B), inside view (I), ocean view (O), or suite (S); and a choice of 2 types of excursions: water sports (W) or horseback riding (H). If you are choosing only one of each, list the sample space in regard to the vacations (combinations of destinations, rooms, and excursions) you could pick from.

2. Which of the following statements is an example of empirical probability?

   o Keith needs to roll five dice of the same number to win the game, and wants to know what the chances of this are.

   o 506 of the 1500 female students in Madison's freshman class are from her home state. Aaron would like to determine the chances of her roommate being from her home state.

   o Aaron spends 5 days catching a total of 12 fish each day and counting how many of the fish he caught are bass. His goal is to determine the percentage of fish in his pond that are bass.

   o Annie is wondering how likely it is that the first card drawn from a standard deck of 52 cards will be red.

3. Which of the following statements is an example of classical probability?

   o Aaron spends 5 days catching a total of 12 fish each day and counting how many of the fish he caught are trout. His goal is to determine the percentage of fish in his pond that are trout.

   o To find the percentage of the time that five of the same number will be rolled when rolling five dice, Cody rolls five dice 50 times each day for 10 days and records the results of each roll.

   o In order to find the percentage of red cards in a large pile of playing cards, Annie draws 50 cards and tallies up the number of black and red cards she sees.

   o 506 of the 1500 female students in Lucia's freshman class are from her home state. Aaron would like to determine the chances of her roommate being from her home state.

4. A sample of 400 full-time employees were surveyed on their feelings about their benefits package. 74 of those studied did not use all of their vacation days last year, yet 88 of those studied expressed a desire for more vacation time. Based on this sample, if a full-time employee is chosen at random, what is the probability that he or she is content with the vacation allowance? Express your answer as a fraction in lowest terms or a decimal rounded to the nearest millionth.

5. A card is drawn from a standard deck of 52 cards. Find the probability that the card is neither red nor black.

6. Jesse has 197 songs on a playlist. He's categorized them in the following manner: 10 hip hop, 33 country, 26 pop, 18 jazz, 39 gospel, 22 rap, and 49 Latin American. If Jesse begins listening to his playlist on shuffle, what is the probability that the first song played is a country song? Express your answer as a fraction in lowest terms or a decimal rounded to the nearest millionth.

7. As students were exiting the student center on campus, Vicki took note whether they were listening to headphones. The table shows results that she collected.

Based on Vicki's data, what is the probability that a randomly selected student will have headphones in their ears when exiting the student center? Express your answer as a fraction in lowest terms or a decimal rounded to the nearest millionth.

Data for Students Exiting the Student Center	
	Number of Students
Headphones	24
No Headphones	50

8. The blood types of 600 people are collected at a doctor's office. The table shows the breakdown of patients per blood type. If a person from this group is selected at random, what is the probability that this person has type A blood? Express your answer as a fraction in lowest terms or a decimal rounded to the nearest millionth.

Blood Type Survey Results	
Blood Type	**Number of Patients**
A	165
B	185
O	195
AB	55

9. An experiment is performed where a 4-color spinner is spun and then another 4-color spinner is spun. The possible outcomes for both events are red (R), blue (B), yellow (Y), and green (G). Identify the sample space for this experiment.

10. What is the probability that a card drawn randomly from a standard deck of 52 cards is a seven? Express your answer as a fraction in lowest terms or a decimal rounded to the nearest millionth.

8.4 Counting Principles: Combinations and Permutations

Tree Diagram

A tree diagram uses branches to indicate _____ at the next state of outcomes.

Fundamental Counting Principle

For a sequence of n experiments where the first experiment has k_1 outcomes, the second experiment has k_2 outcomes, the third experiment has k_3 outcomes, and so forth, the total number of possible outcomes for the sequence of experiments is $(k_1)(k_2)(k_3)\ldots(k_n)$.

Replacement

With replacement: When counting possible outcomes with replacement, objects are placed back into consideration for the following choice.

Without replacement: When counting possible outcomes without replacement, objects are *not* placed back into consideration for the following choice.

n Factorial

In general, $n!$ (read "**n factorial**") is the product of all the positive integers less than or equal to n, where n is a positive integer.

$$n! = n(n-1)(n-2)(n-3)\ldots(2)(1)$$

Note that _____ is defined to be 1.

Example 3: Calculating Factorials

Calculate the value of the following factorial expressions.

a. $8!$

b. $\dfrac{3!}{0!}$

c. $\dfrac{89!}{87!}$

d. $\dfrac{7!}{(5-1)!}$

e. $\dfrac{5!}{3!\,(4-2)!}$

Solution

Combinations

A **combination** involves choosing a specific number of objects from a particular group of objects, using each only once, when the _____ in which they are chosen _____.

Permutations

A **permutation** involves choosing a specific number of objects from a particular group of objects, using each only once, when the _____ in which they are chosen _____.

Combinations and Permutations of n Objects Taken r at a Time

The number of ways to select r objects from a total of n objects is found by the following two formulas. (Note that $r \leq n$.)

When order is not important, use the following formula for a **combination**.

_____

When order is important, use the following formula for a **permutation**.

_____

Example 4: Using Combinations

Let's calculate the number of possibilities for our sandwich example. Suppose there are 18 toppings to choose from once you've decided on bread, meat, and cheese. How many different possible sandwiches are there if you choose 4 different toppings?

Solution

Example 6: Using Permutations

How many possible ways are there to arrange the order of appearance for the contestants in the local talent show, if there are 15 contestants all together?

Solution

Permutations with Repeated Objects

The number of distinguishable _____ of n objects, of which k_1 are all alike, k_2 are all alike, and so forth is given by

$$\frac{n!}{(k_1!)\,(k_2!)\,(k_3!)\ldots(k_p!)},$$

where $k_1 + k_2 + \ldots + k_p = n$.

8.4 Exercises

1. Evaluate the following expression.

7!

2. Evaluate the factorial expression.

$$\frac{19!}{16!\,(4-1)!}$$

3. Evaluate the factorial expression.

$$\frac{22!}{11\,(24-4)!}$$

4. Evaluate the expression. Express your answer as a fraction in lowest terms or a decimal rounded to the nearest thousandth, if necessary.

$$\frac{_5C_4}{_5P_4}$$

5. Lucy was born on 11/11/2000. How many eight-digit codes could she make using the digits in her birthday?

6. There are 12 guests at your home for a dinner party. In how many ways could the guests arrange themselves on a four person couch?

7. John is studying photography, and he was asked to submit 7 photographs from his collection to exhibit at the fair. He has 14 photographs that he thinks are show worthy. In how many ways can the photographs be chosen?

8. The Jacksons are planning their next family night. They always have dinner out somewhere and then do something fun together. There are 2 adults and 3 children in the family. Each family member is allowed 4 meal suggestions, and each child is allowed 2 activity suggestions. Assuming no family members choose the same thing, how many different family night possibilities are there?

9. A mother duck lines her 7 ducklings up behind her. In how many ways can the ducklings line up?

10. Felicia is deciding on her schedule for next semester. She must take each of the following classes: English 102, Spanish 102, History 102, and College Algebra. If there are 16 sections of English 102, 9 sections of Spanish 102, 13 sections of History 102, and 13 sections of College Algebra, how many different possible schedules are there for Felicia to choose from? Assume there are no time conflicts between the different classes.

8.5 Counting Principles and Probability

Complement

The _____ of event E, denoted by E^c, consists of all outcomes in the sample space that are *not* in event E.

Example 3: Finding the Complement of an Event

Describe the complement for each of the following events.

a. Rolling an even number on a die.

b. Choosing a number that doesn't end in 1, from all positive two-digit whole numbers.

c. From a class of 52 students, choosing a student who is over 21 years old.

Solution

Complement Rules of Probability

1. $P(E) + P(E^c) =$ _____

2. $P(E) =$ _____

3. $P(E^c) =$ _____

Example 4: Finding Probability Using Complements

Using the data given, find the following probabilities involving nuts imported into the United States between 2006-2011.

Table 2: US Import Destinations by Weight (in pounds) for Fresh or Dried Walnuts and Pistachios, 2006–2011		
Walnuts	India	4373
	Mexico	1268
	Spain	5239
	China	1533
	Austria	1938
	Other countries	4157
Pistachios	Iran	2012
	Turkey	2030
	Hong Kong	262
	Switzerland	64
	Italy	115
	Other countries	323
Source: USDA. "Fruit and Tree Nut Data." http://www.ers.usda.gov/data-products/fruit-and-tree-nut-data/data-by-commodity.aspx (http://www.ers.usda.gov/data-products/fruit-and-tree-nut-data/data-by-commodity.aspx)		

a. The probability that the walnuts you consumed during this period were from Austria.

b. The probability that the walnuts you consumed during this period were from somewhere other than Austria.

c. Assume that you also purchased pistachios during this time period. What is the probability that the pistachios came from somewhere other than Italy and Switzerland?

Solution

8.5 Exercises

1. There are 4 sets of balls numbered 1 through 9 placed in a bowl. If 4 balls are randomly chosen without replacement, find the probability that the balls have the same number. Express your answer as a fraction in lowest terms or a decimal rounded to the nearest millionth.

2. It is presentation day in class and your instructor is drawing names from a hat to determine the order of the presentations. If there are 16 students in the class, what is the probability that the first 3 presentations will be by Ben, Marcus, and Kenny, in that order? Express your answer as a fraction in lowest terms or a decimal rounded to the nearest millionth.

3. A box of jerseys for a pick-up game of basketball contains 10 extra-large jerseys, 7 large jerseys, and 5 medium jerseys. If you are first to the box and grab 2 jerseys, what is the probability that you randomly grab 2 medium jerseys? Express your answer as a fraction in lowest terms or a decimal rounded to the nearest millionth.

4. Let the event E be the sum of a pair of dice that is divisible by 4. List the events in E^c.

5. Liam and Michael are going to play video games this afternoon. Together, they have 41 video games. If they decide to randomly choose two video games, what is the probability that the two they choose will consist of each of their favorite video games? Assume they have different favorites. Express your answer as a fraction in lowest terms or a decimal rounded to the nearest millionth.

6. Julia sets up a passcode on her tablet, which allows only four-digit codes. A spy sneaks a look at Julia's tablet and sees her fingerprints on the screen over four numbers. What is the probability the spy is able to unlock the tablet on his first try? Express your answer as a fraction in lowest terms or a decimal rounded to the nearest millionth.

7. Kiara sets up a passcode on her smart phone, which allows only seven-digit codes. She has heard that it's more secure if a digit is repeated in the passcode. A spy sneaks a look at Kiara's smart phone and sees her fingerprints on the screen over six numbers. What is the probability the spy is able to unlock the smart phone on his first try? Express your answer as a fraction in lowest terms or a decimal rounded to the nearest millionth.

8. The following table lists the ages and number of players per age group that are on a hockey team. Let $A = \{$hockey players who are 11 years old$\}$. How many players are in the complement of A?

Ages of Players on Hockey Team	
Age	**Number of Players**
8	2
9	2
10	6
11	2

9. Find the probability of choosing a letter other than the letter E from a bag that contains the eighteen letters of the name WILLIAM SHAKESPEARE. Express your answer as a fraction in lowest terms or a decimal rounded to the nearest millionth.

10. Using the table containing the breakdown of all employees on nonfarm payrolls in the United States during March 2014, find the probability that a randomly selected U.S. worker was not in either Retail Trade or Utilities. Express your answer as a fraction in lowest terms or a decimal rounded to the nearest millionth.

Employees on Nonfarm Payrolls (in Thousands), March 2014		
	Area of Employment	Number of Employees (in Thousands)
Private sector	Goods-Producing	18,558.2
	Wholesale Trade	5803.7
	Retail Trade	15,004.0
	Transportation and Warehousing	4524.8
	Utilities	550.3
	Information	2653.0
	Financial Activities	7870.0
	Professional and Business Services	18,832.0
	Education and Health Services	21,481.0
	Leisure and Hospitality	14,143.0
	Other Private Service-Providing Services	5464.0
Public sector	Federal Government	2705.0
	State Government	5217.0
	Local Government	14,341.0
	Total Nonfarm Employees	137,147.0
	Source: Bureau of Labor Statistics. "Table B-1. Employees on nonfarm payrolls by industry sector and selected industry detail." Accessed June 2014. http://www.bls.gov/news.release/empsit.t17.htm	

8.6 Probability Rules and Bayes' Theorem

Addition Rule for Probability

The _____ of Event *A* happening *or* Event *B* happening is

$$P(A \text{ or } B) = P(A) + P(B) - P(A \text{ and } B).$$

Example 2: Applying the Addition Rule for Probability

Recall the example from Section 7.2 where you won a new iPod. You can have any iPod you want, you just have to go to the store and pick it out. We can create a tree diagram to list the possible iPods you could choose from.

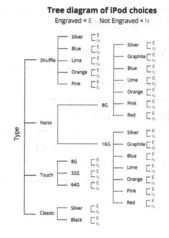

Tree diagram of iPod choices
Engraved = E   Not Engraved = N

Assuming that you are equally likely to choose any of the 48 iPods, what is the probability that the iPod you choose is orange or not engraved?

Solution

Addition Rule for Mutually Exclusive Events

The probability of Event A happening *or* Event B happening when A and B have no outcomes in common is

$$P(A \text{ or } B) = \underline{\hspace{6cm}}.$$

Example 4: Applying the Addition Rule for Mutually Exclusive Events

Choosing a college can be an exciting and nervous time in the life of a high school student. Emma has finally narrowed down her choices to the top 4. She's also given each school a probability based on certain characteristics.

Table 2: University Probabilities		
University	**Characteristic**	**Probability**
A	Closest to home	$P(A) = 0.25$
B	Best sports	$P(B) = 0.10$
C	Her best friend's choice	$P(C) = 0.30$
D	Best academic program of her choice	$P(D) = 0.35$

What is the probability that Emma ends up at University B or University D?

Solution

Independent Events

Independent events are events where the result of one event $\underline{\hspace{5cm}}$ the probability of the other.

Dependent Events

Dependent events are events where the result of one event $\underline{\hspace{3cm}}$ the probability of the other.

Multiplication Rule for Independent Events

A and B are **independent events** when the probability of Event A happening *and* Event B happening is given by $P(A \text{ and } B) =$ _____ .

Example 6: Multiplication Rule for Independent Events

Given a fair die and a standard deck of 52 cards, find the probability of rolling a 6 *and* drawing an ace.

Solution

Example 7: Multiplication Rule for Independent Events

Suppose we know the following breakdown for internal medicine/pediatric hospitalists who work at Madison Regional Hospital and the ages of their patients on a given day.

Table 3: Hospitalists at Madison Regional Hospital		
		Number
Hospitalist Gender	Male	6
	Female	7
Patient Age	< 35	16
	$35–55$	35
	> 55	21

What is the probability that the first patient treated is over 55 years old and treated by a male hospitalist at Madison Regional Hospital?

Solution

Conditional Probability

The **conditional probability** of Event B happening, given Event A, is the probability of Event B assuming that Event A has already, or will at some point, occur. The conditional probability is written _____ , and read *the probability of B, given A.*

Multiplication Rule for Dependent Events

If A and B are **dependent events**, the probability of Event A happening *and* Event B happening is

$$\text{\underline{\hspace{4cm}}} = P(A) \cdot P(B|A).$$

_____

$$P(A|B) = \frac{P(B|A) \cdot P(A)}{P(B)},$$

when $P(B) > 0$.

8.6 Exercises

1. Three cards are drawn with replacement from a standard deck of 52 cards. Find the the probability that the first card will be a heart, the second card will be a red card, and the third card will be a queen. Express your answer as a fraction in lowest terms or a decimal rounded to the nearest millionth.

2. Mr. Griese's philosophy class has 104 students, classified by academic year and gender, as illustrated in the following table. Mr. Griese randomly chooses one student to collect yesterday's work. What is the probability that he selects a female, given that he chooses randomly from only the sophomores? Express your answer as a fraction in lowest terms or a decimal rounded to the nearest millionth.

Mr. Griese's Philosophy Class		
Classification	**Males**	**Females**
Freshmen	16	14
Sophomores	5	13
Juniors	16	11
Seniors	12	17

3. Use the following table to find the probability that a randomly chosen member of the Student Government Board is a graduate student or lives in off-campus housing. Express your answer as a fraction in lowest terms or a decimal rounded to the nearest millionth.

Students on the Student Government Board		
Classification	**On-Campus Housing**	**Off-Campus Housing**
Freshman	1	2
Sophomore	2	2
Junior	3	3
Senior	1	3
Graduate Student	1	1

4. The following table shows the breakdown of opinions for both faculty and students in a recent survey about the new restructuring of the campus to include an Honors College. Find the probability that a randomly selected person is either a student opposed to the change or a faculty member who has an opinion either for or against. Express your answer as a fraction in lowest terms or a decimal rounded to the nearest millionth.

Survey Results on Restructuring Campus to Include an Honors College				
Category	Favor	Oppose	Neutral	Total
Faculty	12	7	10	29
Students	36	74	68	178
Total	48	81	78	207

5. A bag of 13 marbles contains 5 marbles with red on them, 6 with blue on them, 7 with green on them, and 5 with red and green on them. What is the probability that a randomly chosen marble has either green or red on it? Note that these events are not mutually exclusive. Express your answer as a fraction in lowest terms or a decimal rounded to the nearest millionth.

6. Select all of the situations that contain independent events.

   a. Flipping a coin and getting heads twice in a row.

   b. Smoking cigarettes and getting lung cancer.

   c. Eating a candy bar and test results showing a high level of sugar in your bloodstream.

   d. Two fair dice are rolled. The first die lands on five and the second die lands on one.

7. A swim team consists of 6 boys and 7 girls. A relay team of 4 swimmers is chosen at random from the team members. What is the probability that there are 3 boys on the relay team given that there is 1 girl on the relay team? Express your answer as a fraction in lowest terms or a decimal rounded to the nearest millionth.

8. The following is a table showing the results of a poll taken on campus.

Will You Vote in the Upcoming Election?		
	Male	Female
Yes	19	24
No	16	12
Not decided	28	23

What is the probability that a randomly selected student from this poll has **decided not** to vote in the upcoming election? Express your answer as a fraction in lowest terms or a decimal rounded to the nearest millionth.

9. The following is a table showing the results of a poll taken on campus.

Will You Vote in the Upcoming Election?		
	Male	Female
Yes	34	16
No	20	32
Not decided	30	36

What is the probability that a randomly selected student from this poll is **female** or will vote in the upcoming election? Express your answer as a fraction in lowest terms or a decimal rounded to the nearest millionth.

10. A swim team consists of 6 boys and 7 girls. A relay team of 4 swimmers is chosen at random from the team members. What is the probability that there are 3 boys on the relay team given that there is 1 girl on the relay team? Express your answer as a fraction in lowest terms or a decimal rounded to the nearest millionth.

8.7 Expected Value

Expected Value

The formula for **expected value** of an event X is

$$E(X) = x_1 P(x_1) + x_2 P(x_2) + x_3 P(x_3) + \cdots + x_n P(x_n),$$

where x_i is the ith outcome and $P(x_i)$ is the probability of x_i.

The Sum of Expected Values

The expected value of the union of two mutually exclusive events, X and Y, is equal to the sum of the individual expected values. Formally, we can write this as

$$E(X \cup Y) = \underline{\hspace{4cm}}.$$

Example 2: Expected Winnings

In American roulette, the wheel contains the numbers 1 through 36, alternating between black and red. There are two green spaces numbered 0 and 00.

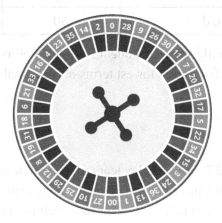

a. Calculate the probability of the roulette ball landing on a red pocket (or black for that matter, since the number of red and black pockets is the same).

b. Calculate the probability of the ball not landing on a red pocket.

c. A player bets $1.00 on red to play the game. Calculate his expected winnings.

Solution

8.7 Exercises

1. An insurance company claims the probability of surviving a certain type of cancer is 20%. What are the odds of surviving? Express your answer in the form $a : b$.

2. Suppose the odds for a bet are $12 : 1$. Your friend tells you that he thinks the odds are too generous. Select all of the odds that are less generous.

    ○ $19 : 1$

    ○ $7 : 1$

    ○ $17 : 1$

    ○ $14 : 1$

3. During the NCAA basketball tournament season, affectionately called *March Madness*, part of one team's strategy is to foul their opponent if his free-throw shooting percentage is lower than his two-point field goal percentage. Juan's free-throw shooting percentage is lower and is only 57.4%. After being fouled he gets two free-throw shots each worth one point. Calculate the expected value of the number of points Juan makes when he shoots two free-throw shots.

4. You need to borrow money for gas, so you ask your mother and your sister. You can only borrow money from one of them. Before giving you money, they each say they will make you play a game. Your sister says she wants you to spin a spinner with six outcomes, numbered 1 through 6, on it. She will give you $5 times the number that the spinner lands on. Your mother says she wants you to flip a fair coin. She will give you $12 for heads and $28 for tails. Determine the expected value of each game and decide which offer you should take.

5. Suppose that 15% of the time Todd goes swimming twice a month, 30% of the time he goes swimming once a month, and 55% of the time he doesn't go swimming at all in a given month. What is the expected value for the number of times Todd goes swimming during a month?

6. Suppose the probability that it rains tomorrow is 0.96. What are the odds of it raining? Express your answer in the form $a : b$.

7. Calculate the expected value of the scenario.

x_i	$P(x_i)$
−$1.25	0.2
−$0.25	0.3
$0	0.2
$0.25	0.3

8. Calculate the expected value of the scenario.

x_i	$P(x_i)$
1	0.27
2	0.46
3	0.02
4	0.22
5	0.03

9. Suppose that you and a friend are playing cards and decide to make a bet. If you draw three black cards in succession from a standard deck of 52 cards without replacement, you win $50. Otherwise, you pay your friend $20. If the same bet was made 25 times, how much would you expect to win or lose? Round your answer to the nearest cent, if necessary.

10. Suppose that you and a friend are playing cards and decide to make a bet. If your friend draws three face cards, where a face card is a Jack, a Queen, or a King, in succession from a standard deck of 52 cards without replacement, you give him $40. Otherwise, he pays you $10. What is the expected value of your bet? Round your answer to the nearest cent, if necessary.

CHAPTER 9

Statistics

9.1 Collecting Data

Population

A **population** is the particular _____ in a study.

Data

Data is _____ collected for a study.

Census

A **census** involves collecting data from _____ of the population.

Parameter

A **parameter** is the _____ of a particular population characteristic.

Sample

A **sample** is a _____ of a population.

Sample Statistic

A **sample statistic** is the numerical description of a _____ of a sample.

Example 1: Identifying Parts of a Survey

Two shortened survey reports are given. In each report, identify the following: the population, the sample, the results, and whether the results represent a sample statistic or a population parameter.

a. A headline about the rising obesity among young people led a school board to survey local high school students. Out of 231 students surveyed, 58% reported eating a "high fat" snack at least 4 times a week.

b. A nonprofit organization interviewed 618 adult shoppers at malls across Louisiana about their views on obesity in youths. The resulting report stated that an estimated 48% of Louisiana adults are in favor of government regulation of "high fat" fast food options.

Solution

Representative Sample

A **representative sample** is one that has the same _____ as the population and does not favor one group of the population over another.

Random Sample

A **random sample** is one in which _____ of the population has an equal chance of being selected.

Stratified Sample

A **stratified sample** is one in which members of the population are divided into two or more _____, called strata, that share similar characteristics like age, gender, or ethnicity. A random sample from *each* stratum is then drawn.

Cluster Sample

A **cluster sample** is one chosen by dividing the population into groups, called clusters, that are each _____. The researcher then randomly selects some of the clusters. The sample consists of the data collected from *every* member of each cluster selected.

Systematic Sample

A **systematic sample** is one chosen by selecting _____ of the population.

Convenience Sample

A **convenience sample** is one in which the sample is "convenient" to select. It is so named because it is convenient for the _____ .

Example 2: Identifying Sampling Techniques

Identify the sampling technique used to obtain a sample in each of the following situations.

a. To conduct a survey on collegiate social life, you knock on every 5th dorm room door on campus.

b. Student ID numbers are randomly selected from a computer print out for free tickets to the championship game.

c. Fourth grade reading levels across the county were analyzed by the school board by randomly selecting 25 fourth graders from each school in the county district.

d. In order to determine what ice cream flavors would sell best, a grocery store polls shoppers that are in the frozen foods section.

e. To determine the average number of cars per household, each household in 4 of the 20 local counties were sent a survey regarding car ownership.

Solution

9.1 Exercises

1. Identify the term that best matches the given definition.

A sample in which members of the population are divided into groups that are similar to the population. Data is collected from every member of the selected groups.

2. Four scenarios of statistical studies are given below. Decide which study uses a population parameter.

   ○ 15% of the attendees at a town council meeting in a rural area reported not using the internet on a daily basis.

   ○ 1000 of the 1883 participants in last year's Ironman in Kona, Hawaii were interviewed. 68% of those interviewed said they became interested in triathlons after first competing in marathons.

   ○ As part of her senior project, Kaisa asked 84 of her classmates if they voted in the last presidential election. 39 said yes.

   ○ The National Institute of Pediatric Dentistry reports that 42% of children ages 2-11 have had at least one cavity.

3. Identify the population, the sample, and any population parameters or sample statistics in the given scenario.

   In the 1960s, a poll was taken of 2617 homeowners in the United States. The average price of homes owned by those surveyed was $18,500.

4. Identify the sampling technique used in the given scenario.

   The English department is interested in overall performance in its entry-level 101 course. The department chair randomly selects 10 students' final grades from each of the 30 sections offered and calculates the average final grade.

5. Identify the appropriate sampling method for the following scenario. Be sure to consider potential biases that the researcher might need to consider.

   A politician wants to survey his constituents on key issues. His voting district contains five neighborhoods of generally different age, ethnicity, and income. The issues affect all voters, and no single neighborhood contains a representative sample of the district.

6. Identify the type of sample.

   A sample that is chosen by selecting every n^{th} member of the population.

7. Identify the type of sample.

A sample in which every member of the population has an equal chance of being selected.

8. Identify the population in the given scenario.

Fit and Healthy magazine surveyed 1689 people. The resulting report stated that an estimated 56 % of the U.S. population are self-conscious about their weight.

9. Identify the sample in the given scenario.

In the 1960 s, a poll was taken of 2617 homeowners in the United States. The average price of homes owned by those surveyed was $ 18,500.

10. Explain how cluster sampling is different from stratified sampling.

9.2 Displaying Data

Frequency Distribution

A **frequency distribution** is a count of every member of the data set and _____ each value occurs, that is, the frequency of the data value.

Example 1: Interpreting a Frequency Distribution

The following frequency distribution shows the number of crashes with roadside objects along State Route 3 in the state of Washington. After looking over the table, answer the questions that follow.

Table 1: Crashes with Roadside Objects Along State Route 3, Washington	
Roadside Object	**Number of Crashes (% of total)**
Guardrail	57 (15.36)
Earth Bank	55 (14.82)
Ditch	42 (11.32)
Tree	42 (11.32)
Concrete Barrier	38 (10.24)
Over Embankment	31 (8.36)
Utility Pole	20 (5.39)
Wood Sign Support	19 (5.12)
Bridge Rail	17 (4.58)
Culvert	7 (1.89)
Boulder	6 (1.62)
Luminaire	6 (1.62)
Mailbox	5 (1.35)
Fence	5 (1.35)
Building	5 (1.35)
Other Object	16 (4.31)

Source: Jinsun Lee and Fred Manning. "Analysis of Roadside Accident Frequency and Severity and Roadside Safety Management." Washington State Transportation Commission. December 1999. http://www.wsdot.wa.gov/research/reports/fullreports/475.1.pdf

a. What do the numbers in the parentheses represent?

b. Which roadside object was involved in the most crashes?

c. How many total crashes did the survey cover?

Solution

Example 4: Interpreting a Frequency Distribution

The San Francisco Bay Area surveyed their residents about the nearly 17 million intraregional daily trips the population makes. Along with other data that they collected, they asked people to identify where their trips originated and how long they traveled. Table 7 organizes the travel data by trip purpose and travel time. The trip purposes recorded in the survey were work, shop, social/recreational, and school. Home-based trips were those that either started or ended at the residence of the trip maker. Non-home based trips were those that neither started nor ended at the residence.

Answer the following questions about the frequency chart and its data.

Travel Time (in Minutes)	Home-Based Work	Home-Based Shop (Other)	Home-Based Social/Rec.	Home-Based School	Non-Home Based	Total Purposes
0 – 5.0	269,350	807,952	272,973	242,906	953,920	2,547,101
5.1 – 10.0	416,075	881,203	357,196	282,305	917,191	2,853,970
10.1 – 15.0	886,859	1,217,440	491,636	434,810	1,206,126	4,236,871
15.1 – 20.0	440,776	316,201	137,957	139,623	331,661	1,366,218
20.1 – 25.0	283,777	200,142	84,876	103,114	213,205	885,114
25.1 – 30.0	807,327	391,393	228,241	211,851	455,129	2,093,941
30.1 – 35.0	155,706	53,270	27,501	40,179	71,154	347,810
35.1 – 40.0	169,675	50,564	34,060	38,579	79,675	372,553
40.1 – 45.0	327,028	101,392	60,827	69,478	150,929	709,654
45.1 – 50.0	88,325	25,253	16,074	11,509	30,240	171,401
50.1 – 55.0	56,228	15,686	10,876	13,804	25,282	121,876
55.1 – 60.0	227,844	76,162	48,138	38,221	94,638	485,003
60.1 – 65.0	45,851	10,352	5287	3284	16,144	80,918
65.1 – 70.0	42,824	11,974	7697	3952	14,434	80,881
70.1 – 75.0	81,350	20,182	14,642	10,800	35,675	162,649
75.1 – 80.0	22,123	5876	3311	3321	9344	43,975
80.1 – 85.0	13,594	3852	3121	2495	4950	28,012
85.1 – 90.0	50,220	13,375	12,996	4947	27,684	109,222
90+	68,457	29,496	26,402	9446	47,242	181,043
Total	4,453,389	4,231,765	1,843,811	1,664,624	4,684,623	16,878,212

Table 7: Number of Intraregional Trips

Source: National Transportation Library. "San Francisco Bay Area 1990 Regional Travel Characteristics - WP #4 - MTC Travel Survey." http://ntl.bts.gov/DOCS/SF.html

a. How many classes are there in the frequency distribution?

b. Find the class width. (Exclude the first and last classes.)

c. Which guideline is ignored in this published frequency distribution? What sorts of questions arise from this inconsistency?

d. Which class contains the largest number of school-related trips, and what is its frequency?

e. Is the statement "Most people take 10 – 15 minutes to travel to school in the San Francisco Bay Area" a true statement?

Solution

Pie Chart

A **pie chart** shows how large each category is in relation to _____ .

Bar Graph

A **bar graph** is a chart of _____ in which the height of the bar represents the amount of data in each category.

Example 8: Reading Graphs

Considering the map of the distribution of the slave population of the Southern United States in Figure 8 and the data in Table 9, answer the following questions. The darker the shading is on the map, the higher the slave population was in that area.

a. Which states had a higher slave population than free population in the 1860 census?

b. Which state had the highest number of slaves in 1860?

c. One of the reasons that made the map so popular was the visual account of slavery by shading. Which areas had the heaviest concentration of slaves?

Solution

9.2 Exercises

1. The grades on the second statistics test for Mr. Montgomery's class are in the following list. Complete the frequency distribution table for the grades.

D, D, C, C, D, D, D, C, C, F, A, C, A, C, C, F, F, D, D, C, B

2. The following table represents a grouped frequency distribution of the number of hours spent on the computer per week for 55 students. What is the class width?

Hours	Number of Students
0.0–3.4	5
3.5–6.9	16
7.0–10.4	15
10.5–13.9	13
14.0–17.4	6

3. The following table represents a grouped frequency distribution of the number of hours spent on the computer per week for 53 students. What is the value of the upper class limit of the 4th class?

Hours	Number of Students
0.0–3.4	2
3.5–6.9	16
7.0–10.4	15
10.5–13.9	14
14.0–17.4	6

4. The following table represents a grouped frequency distribution of the number of hours spent on the computer per week for 51 students. What percentage of students used the computer between 7 and 10.4 hours per week? Round your answer to one decimal place, if necessary.

Hours	Number of Students
0.0–3.4	3
3.5–6.9	16
7.0–10.4	15
10.5–13.9	12
14.0–17.4	5

5. The following table represents a grouped frequency distribution of the number of hours spent on the computer per week for 47 students. What percentage of students used the computer more than 13.4 hours per week? Round your answer to the nearest tenth, if necessary.

Hours	Number of Students
0.0–4.4	3
4.5–8.9	16
9.0–13.4	13
13.5–17.9	11
18.0–22.4	4

6. The following pie chart represents the destinations for recent biology graduates in the US. Use the chart to determine which destination has a similar percent of graduates to that of microbiology.

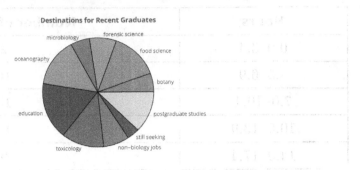

7. The following bar graph shows the profits or losses for a small retail store in its first year of operation. Use the graph to determine the number of months in which the store had a loss.

8. The following line graph shows the price of a stock over the previous two weeks. Use the graph to determine the number of days in which the price was between $50 and $60.

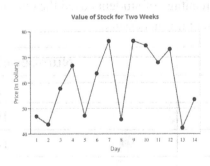

9. Determine how the following graph is misleading, if at all.

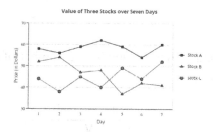

10. Fill in the blank with the correct term.

A_____ is best to use when showing data over a time period.

9.3 Describing and Analyzing Data

Arithmetic Mean

The **mean** is the sum of all of the data values divided by the number of data points. Formally, the formula for the _____ **mean** is

$$\mu = \frac{x_1 + x_2 + \ldots + x_N}{N}.$$

The formula for the _____ **mean** is

$$\overline{x} = \frac{x_1 + x_2 + \ldots + x_n}{n}.$$

x_i is the ith data value, N is the number of data values in the population, and n is the number of data values in the sample.

Median

The **median** of a data set is the middle value in an _____ array of the data.

Mode

The **mode** is the value in the data set that _____ .

Example 3: Finding the Mode

Find the mode of each of the following sets of data. State if the data set is unimodal, bimodal, multimodal, or has no mode.

a. Preferred color of cell phone cases among students

lemon, gunmetal, violet, turquoise, lime, violet, lemon, orange, red, lemon, pink, violet, lime, violet, lemon, pink, gunmetal, red, turquoise, violet, violet, gunmetal, turquoise, red, violet, turquoise, orange, pink, violet, violet, turquoise, violet, pink

b. Favorite football jersey number

32, 18, 99, 12, 7, 10, 28, 56, 13, 16, 19, 51, 23, 78

c. Ages of children at the community playground one afternoon

12, 4, 2, 7, 8, 4, 10, 6, 5, 7, 7, 4, 3

d. Number of ATM withdrawals per hour at the downtown branch of University Bank

$$10, 13, 9, 13, 9, 14, 10, 14$$

Solution

Outlier

An **outlier** is a data value that is _____ compared with the rest of the data values in the set.

Range

The **range** is the _____ between the largest and smallest values in the data set, which tells you the distance covered on the number line between the two extremes.

$$\text{range} = \text{maximum data value} - \text{minimum data value}$$

Example 5: Finding the Range

Find the range of the following sets of data.

a. The number of students enrolled as computer science majors over the past 12 semesters

$$5, 21, 54, 33, 12, 14, 36, 40, 27, 29, 37, 22$$

b. The number of shoppers at a gas station downtown Monday through Sunday one week

$$1007, 1010, 1006, 1005, 1054, 1021, 1005$$

Solution

Standard Deviation

The **standard deviation** is a measure of how much we might expect a member of the data set to

_____.

The formula for finding the population standard deviation is

$$\sigma = \sqrt{\frac{\sum (x_i - \mu)^2}{N}}$$

where x_i is the ith data value, μ is the population mean, and N is the size of the population.

For a sample, the standard deviation is

$$s = \sqrt{\frac{\sum (x_i - \overline{x})^2}{n - 1}}$$

where x_i is the ith data value, $\overline{x}$ is the sample mean, and n is the sample size.

Percentile

Percentiles divide the data into _____ parts and tell you approximately what percentage of the data lies at or below a given value.

Example 9: Interpreting Percentiles

Sierra received her scores from taking a mathematics placement test for her chosen university. Choose the best explanation for what it means for her to be in the 61st percentile.

a. She correctly answered 61% of the answers on the test.

b. 61% of people taking the test scored the same as Sierra.

c. Sierra's score was at least as good as 61% of the people taking the test.

d. Sierra missed 39% of the test questions.

Solution

Quartiles

Q_1 =**First Quartile**= _____ percentile, that is, _____ of the data is less than or equal to this value.

Q_2 =**Second Quartile**= _____ percentile, that is, _____ of the data is less than or equal to this value.

Q_3 =**Third Quartile**= _____ percentile, that is, _____ of the data is less than or equal to this value.

By definition, Q_2 will be the same as the median.

Example 10: Interpreting Quartiles

On Karl's recent standardized test results, the picture graph of his score showed he was above the third quartile in language arts. His classmate, Asher, said his score was at the 70th percentile, while Rylie said hers was at the 79th percentile. Which of the three had the best language arts test score?

Solution

9.3 Exercises

1. Find the median of the following data set. Assume the data set is a sample.

   $16, 23, 24, 29, 12, 24, 25, 18, 22, 19, 19$

2. Identify whether the following data set is unimodal, bimodal, multimodal, or has no mode. Assume the data set is a sample and find the modes, if any exist.

   $26, 39, 22, 28, 32, 29, 29, 35, 34, 34, 33$

3. Find the standard deviation of the following data set. Assume the data set is a sample. Round your answer to the nearest hundredth, if necessary.

$23, 46, 19, 25, 41, 26, 26, 32, 31, 31, 44$

4. Identify whether the following data set is unimodal, bimodal, multimodal, or has no mode. Assume the data set is a sample and find the modes, if any exist.

$41, 54, 37, 43, 48, 44, 46, 50, 43, 36, 43$

5. A researcher randomly purchases several different kits of a popular building toy. The following table shows the number of pieces in each kit in the sample. Find the mean of the data. Round your answer to the nearest hundredth, if necessary.

Building Toy Pieces	
171	56
276	361
361	56
115	307
276	120
181	219

6. Frank knows that his first four test grades were 87, 94, 90, and 85. Use the formula $\bar{x} = \dfrac{x_1 + x_2 + \ldots + x_n}{n}$ to find Frank's grade on the fifth test if his test average is 86.8.

7. Consider the following data set.

The bands who have played at a single music venue over the past five years.

Would you be more interested in looking at the mean, median, or mode? State your reasoning.

8. Suppose that IQ scores have a bell-shaped distribution with a mean of 101 and a standard deviation of 14. Using the empirical rule, what percentage of IQ scores are at least 115? Please do not round your answer.

9. Suppose that IQ scores have a bell-shaped distribution with a mean of 95 and a standard deviation of 16. Describe where the highest and lowest 0.3% of IQ scores lie.

10. Identify the term that is defined in the given statement.

The middle value in an ordered array of the data in a data set.

9.4 The Binomial Distribution

Binomial Experiment

A **binomial experiment** is a _____ which satisfies all of the following conditions.

1. There are only two outcomes in each trial of the experiment. (One of the outcomes is usually referred to as a *success*, and the other as a *failure*.)

2. The experiment consists of n identical trials as described in Condition 1.

3. The probability of success on any one trial is denoted by p and does not change from trial to trial. (Note that the probability of a failure is $(1 - p)$ and also does not change from trial to trial.)

4. The trials are independent.

5. The binomial random variable is the count of the number of successes in n trials.

Example 1

Toss a coin 4 times and record the number of heads. Is the number of heads in 4 tosses a binomial random variable?

Solution

Binomial Probability Distribution Function

The **binomial probability distribution function** is

$$P(X = x) = {}_nC_x p^x (1-p)^{n-x}$$

where ${}_nC_x$ represents the number of possible combinations of n objects taken x at a time (without replacement) and is given by

$${}_nC_x = \frac{n!}{x!\,(n-x)!} \text{ where } n! = n\,(n-1)\,(n-2)\cdots(2)\,(1) \text{ and } 0! = 1;$$

$n = $ _____ ,

$p = $ the _____ , and

$x = $ _____ .

Example 3

The US Land Management Office regularly holds a lottery for the lease of government lands. Your company has won the rights to 12 leases. Historically, about 10% of these lands possess sufficient oil reserves for profitable operation. Construct the distribution for the number of leases that will be profitable. What is the probability that at least one of the leases will be profitable?

Solution

Expected Value of a Binomial Random Variable

The **expected value** of a binomial random variable can be computed using the expression

$$\mu = E(X) = \underline{\hspace{2cm}}.$$

where n and p are the parameters of the binomial distribution.

Variance and Standard Deviation of a Binomial Random Variable

To find the **variance** of a binomial random variable, use the expression

$$\sigma^2 = V(X) = \underline{\hspace{4cm}}.$$

Thus, the **standard deviation** of a binomial random variable is given by

$$\sigma = \sqrt{V(X)} = \underline{\hspace{4cm}}.$$

Example 5

Compute the expected value and the variance of the number of profitable leases in Example 3.

Solution

9.4 Exercises

1. A student takes an exam containing 13 multiple choice questions. The probability of choosing a correct answer by knowledgeable guessing is 0.3. If the student makes knowledgeable guesses, what is the probability that he will get exactly 10 questions right? Round your answer to four decimal places.

2. A student takes an exam containing 17 true or false questions. At least 11 correct answers are required to pass. If the student guesses, what is the probability that he will pass? Round your answer to four decimal places.

3. A researcher wishes to conduct a study of the color preferences of new car buyers. Suppose that 40% of this population prefers the color red. If 15 buyers are randomly selected, what is the probability that exactly two-fifths of the buyers would prefer red? Round your answer to four decimal places.

4. A quality control inspector has drawn a sample of 19 light bulbs from a recent production lot. Suppose 30% of the bulbs in the lot are defective. What is the probability that less than 13 but more than 9 bulbs from the sample are defective? Round your answer to four decimal places.

5. The Magazine Mass Marketing Company has received 16 entries in its latest sweepstakes. They know that the probability of receiving a magazine subscription order with an entry form is 0.4. What is the probability that no more than one of the entry forms will include an order? Round your answer to four decimal places.

6. The Magazine Mass Marketing Company has received 14 entries in its latest sweepstakes. They know that the probability of receiving a magazine subscription order with an entry form is 0.6. What is the probability that between 6 and 9 (both inclusive) entry forms include an order? Round your answer to four decimal places.

7. A fair die will be rolled 10 times. What is the probability that an even number is rolled exactly 5 times? Round your answer to four decimal places.

8. A fair die will be rolled 10 times. What is the probability that 4, 5, or 6 is rolled at most 3 times? Round your answer to four decimal places.

9. Parents have always wondered about the sex of a child before it is born. Suppose that the probability of having a male child was 0.5, and that the sex of one child is independent of the sex of other children. What is the probability of having exactly 1 boy out of 5 children? Round your answer to four decimal places.

10. Determine whether or not the given procedure results in a binomial distribution. If not, identify which condition is not met.

Surveying 28 people to determine which statistics courses they have taken.

9.5 The Normal Distribution

Characteristics of the Normal Distribution

a. It is bell-shaped, meaning it has only one _____ that is at the center, and symmetrical.

b. The mean, median, and mode are all the _____ .

c. The total area under the curve is equal to _____ .

d. It is completely defined by its mean and _____ .

z-Score

The **z-score** tells how many standard deviations a particular piece of data lies away from the mean in a normal distribution. The formula for finding a z-score is

$$z = \underline{\hspace{6cm}},$$

or more formally with symbols,

$$z = \frac{x - \mu}{\sigma} \text{ for } \underline{\hspace{3cm}} \qquad \text{and } z = \frac{x - \overline{x}}{s} \text{ for } \underline{\hspace{3cm}} .$$

Example 1: Calculating z-scores

Given that the heights of Canadian women are normally distributed with a mean of 159.5 cm and a standard deviation of 7.1 cm, calculate the z-scores for the following pieces of data.

a. A height of 148.2 cm

b. A height of 160.3 cm

c. A height of 1.7 m

Solution

9.5 Exercises

1. Calculate the standard score of the given x value, $x = 66.6$, where $\bar{x} = 65.6$, $s = 4.7$. Round your answer to two decimal places.

2. Scores on a test have a mean of 66 and a standard deviation of 8. Michael has a score of **73**. Convert Michael's score to a z-score, rounded to the nearest hundredth.

3. The annual rainfall in a town has a mean of **53.63** inches and a standard deviation of **9.93** inches. Last year there was rainfall of **62** inches. How many standard deviations away from the mean is that? Round your answer to two decimal places.

4. A water bottling facility has a mean bottling rate of **32.4** thousand bottles per hour with a standard deviation of **1.84** thousand bottles. A nearby cola bottling facility has a mean bottling rate of **27.5** thousand bottles per hour with a standard deviation of **1.46** thousand bottles. One Wednesday from noon to 1:00 p.m., the water bottling facility bottled **35.1** thousand bottles of water, and the cola bottling facility bottled **29.3** thousand bottles of cola. Which facility increased their efficiency more during that hour?

5. Use the formula for finding a z-score to determine the missing value in the following table. Round your answer to two decimal places, if necessary.

z	x	μ	σ
-2.69	26.40	43.5	?

6. What percentage of the area under the normal curve is to the left of the following z-score? Round your answer to two decimal places.

$z = -1.36$

7. What percentage of the area under the normal curve is to the left of z_1 and to the right of z_2? Round your answer to two decimal places.

$z_1 = 0.10$

$z_2 = 1.21$

8. Last year, the revenue for medical equipment companies had a mean of 60 million dollars with a standard deviation of 15 million. Find the percentage of companies with revenue greater than 41 million dollars. Assume that the distribution is normal. Round your answer to the nearest hundredth.

9. Last year, the revenue for medical equipment companies had a mean of 80 million dollars with a standard deviation of 15 million. Find the percentage of companies with revenue between 53 million and 106 million dollars. Assume that the distribution is normal. Round your answer to the nearest hundredth.

10. What is the minimum z-score that a data point could have in order to remain in the top 25% of a set of data? Round your answer to two decimal places.

9.6 Normal Approximation to the Binomial Distribution

Continuity Correction

Continuity correction is used when a _____ distribution is approximated using a _____ distribution. To apply continuity correction, subtract or add 0.5 (depending on the question at hand) to a selected value in order to find the desired probability.

Example 2

An advertising agency hired on behalf of Tech's development office conducted an ad campaign aimed at making alumni aware of their new capital campaign. Upon completion of the new campaign, the agency claimed that 20% of alumni in the state of Virginia were aware of the new campaign. To validate the claim of the agency, the development office surveyed 1000 alumni in the state and found that 150 were aware of the campaign. Assuming that the ad agency's claim is true, what is the probability that no more than 150 of the alumni in the random sample were aware of the new campaign?

Solution

Example 3

A popular restaurant near Tech's campus accepts 200 reservations on Saturdays, the day of a Tech football game. Given that many of the reservations are made weeks in advance of game day, the restaurant expects that about eight percent will be no-shows. What is the probability that the restaurant will have no more than 20 no-shows on the next Saturday of a football weekend?

Solution

9.6 Exercises

1. Consider the probability that at least 81 out of 299 people have been in a car accident.

Describe the area under the normal curve that would be used to approximate binomial probability.

2. Consider the probability that more than 59 out of 531 software users will call technical support.

Describe the area under the normal curve that would be used to approximate binomial probability.

3. Consider the probability that no more than 19 out of 370 CDs will be defective.

Describe the area under the normal curve that would be used to approximate binomial probability.

4. Consider the probability that less than 95 out of 307 computers will crash in a day.

Describe the area under the normal curve that would be used to approximate binomial probability.

5. Consider the probability that exactly 19 out of 317 registered voters will not vote in the presidential election.

Describe the area under the normal curve that would be used to approximate binomial probability.

6. Consider the probability that fewer than 75 out of 139 registered voters will vote in the presidential election. Assume the probability that a given registered voter will vote in the presidential election is 57%.

Specify whether the normal curve can be used as an approximation to the binomial probability by verifying the necessary conditions.

7. Consider the probability that less than 17 out of 146 houses will lose power once a year. Assume the probability that a given house will lose power once a year is 12%.

Approximate the probability using the normal distribution. Round your answer to four decimal places.

8. Consider the probability that no more than 18 out of 148 cell phone calls will be disconnected. Assume the probability that a given cell phone call will be disconnected is 12%.

   Approximate the probability using the normal distribution. Round your answer to four decimal places.

9. Consider the probability that no less than 94 out of 145 flights will be on-time. Assume the probability that a given flight will be on-time is 61%.

   Approximate the probability using the normal distribution. Round your answer to four decimal places.

10. Consider the probability that exactly 90 out of 155 people will not get the flu this winter. Assume the probability that a given person will not get the flu this winter is 66%.

    Approximate the probability using the normal distribution. Round your answer to four decimal places.

CHAPTER 10

Limits and the Derivative

10.1 One-Sided Limits

One-Sided Limits Defined Informally

1. Left-hand Limits

If the values of a function $y = f(x)$ get closer and closer to some number K as values of x, that are smaller than some number a, get closer and closer to a, then we say that K is the _____ of $f(x)$ **as x approaches a from the** _____. We write

_____.

2. Right-hand Limits

If the values of a function $y = f(x)$ get closer and closer to some number M as values of x, that are larger than some number a, get closer and closer to a, then we say that M is the _____ of $f(x)$ **as x approaches a from the** _____. We write

_____.

Example 1: Finding One-Sided Limits

The graph of the function $f(x) = \begin{cases} x^2 & \text{if } 0 \le x \le 2 \\ 2 & \text{if } 2 < x < 4 \\ x & \text{if } 4 \le x \le 6 \end{cases}$ is shown in the figure below. Find the following one-sided limits by inspecting the graph.

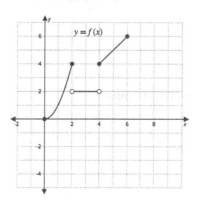

a. $\lim\limits_{x \to 2^-} f(x)$

b. $\lim\limits_{x \to 2^+} f(x)$

c. $\lim\limits_{x \to 4^-} f(x)$

d. $\lim\limits_{x \to 4^+} f(x)$

e. $\lim\limits_{x \to 0^+} f(x)$

Solution

Polynomial Function

A _____ **function** is a function of the form

$$p(x) = a_n x^n + a_{n-1} x^{n-1} + \ldots + a_2 x^2 + a_1 x + a_0$$

where each exponent is a positive integer and $a_n, a_{n-1}, \ldots, a_0$ are real numbers.

One-Sided Limits of Polynomial Functions

If p is a polynomial function and a is a real number, then

$$\lim\limits_{x \to a^-} p(x) = \underline{\quad\quad} \quad \text{and} \quad \lim\limits_{x \to a^+} p(x) = \underline{\quad\quad}.$$

Example 2: Finding One-Sided Limits

Find the values of each of the following one-sided limits.

a. $\lim\limits_{x \to 3^-} (x^2 + 3x - 7)$

b. $\lim\limits_{x \to 2^+} (x^3 - 5x)$

c. $\lim\limits_{x \to 0^-} (4x^2 - 8x + 3)$

Solution

Infinite One-Sided Limits (Unbounded One-Sided Limits)

1. $+\infty$ Limits

If $f(x)$ _____ without bound as x approaches a from the left (or from the right), then we say that $f(x)$ approaches _____ , $+\infty$. We write

a. $\lim\limits_{x \to a^-} f(x) =$ _____

or

b. $\lim\limits_{x \to a^+} f(x) =$ _____ .

2. $-\infty$ Limits

If $f(x)$ _____ without bound as x approaches a from the left (or from the right), then we say that $f(x)$ approaches _____ , $-\infty$. We write

a. $\lim\limits_{x \to a^-} f(x) =$ _____

or

b. $\lim\limits_{x \to a^+} f(x) =$ _____ .

Example 3: Finding Infinite One-Sided Limits

a. Find $\lim\limits_{x \to -3^-} \left(\dfrac{x+5}{x+3} \right)$.

Solution

b. Find $\lim\limits_{x \to -3^+} \left(\dfrac{x+5}{x+3} \right)$.

Solution

Example 4: Finding One-Sided Limits Using a Graph

Study the graph shown for $y = f(x)$ and find the following one-sided limits.

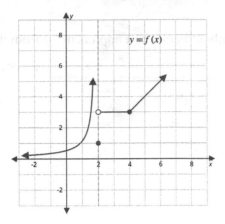

a. $\lim\limits_{x \to 2^-} f(x)$

b. $\lim\limits_{x \to 2^+} f(x)$

c. $\lim\limits_{x \to 4^-} f(x)$

d. $\lim\limits_{x \to 4^+} f(x)$

Solution

10.1 Exercises

1. Consider the following one-sided limit.

$$\lim_{x \to 0^-} \left(\frac{\sqrt{64 + x}}{x} \right)$$

   a. Approximate the limit by filling in the table. Round to the nearest thousandth.

x	y
−1	
−0.1	
−0.01	
−0.001	

   b. Determine the value of the one-sided limit.

2. Use the graph of $y = f(x)$ to find the limits:

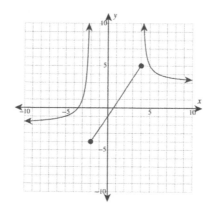

   a. Find $\lim\limits_{x \to -2^-} f(x)$.

b. Find $\lim\limits_{x \to -2^+} f(x)$.

c. Find $\lim\limits_{x \to 4^-} f(x)$.

d. Find $\lim\limits_{x \to 4^+} f(x)$.

3. Find the one-sided limit.

$$\lim\limits_{x \to 0^+} \left(\frac{x - 76}{x} \right)$$

4. Find the one-sided limit.

$$\lim\limits_{x \to 7^-} \left(\frac{x + 6}{x - 7} \right)$$

5. Find the one-sided limits of $f(x)$.

$$f(x) = \begin{cases} 6x + 6 & \text{if } x < -6 \\ -7x^2 - 8 & \text{if } x \geq -6 \end{cases}$$

a. Find $\lim\limits_{x \to -6^-} f(x)$.

b. Find $\lim\limits_{x \to -6^+} f(x)$.

10.2 Limits

Limit Defined Informally

If the values of $f(x)$ get _____ to some number L as values of x that are

_____ than some number t and values of x that are _____ than t get closer and closer to

t (but not equal to t), then **L is the** _____ **of $f(x)$ as x approaches t.**

Limit Defined Symbolically

For a function f and a real number t, if $\lim\limits_{x \to t^-} f(x) = L$ and $\lim\limits_{x \to t^+} f(x) = L$, where L is a real number, then

_____ .

Existence of a Limit

1. We say that the limit of a function f as x approaches a real number _____ if and only if

$\lim\limits_{x \to t} f(x) = L$. where L is a real number.

2. If $\lim\limits_{x \to t^-} f(x) \neq \lim\limits_{x \to t^+} f(x)$, then $\lim\limits_{x \to t} f(x)$ _____ .

3. If $\lim\limits_{x \to t} f(x) = +\infty$ or $\lim\limits_{x \to t} f(x) = -\infty$, then $\lim\limits_{x \to t} f(x)$ _____ .

Example 1: Finding a Limit Using a Graph

Use the graph of $y = f(x)$ in the figure to find

a. $\lim\limits_{x \to 3} f(x)$, and

b. $\lim\limits_{x \to 0} f(x)$.

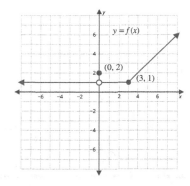

Solution

═══════════════════════════════════════

Example 2: Finding a Limit Using a Graph

Use the graph of $y = g(x)$ in the figure to find

a. $\lim_{x \to -2} g(x)$, and

b. $\lim_{x \to 2} g(x)$, if the limits exist.

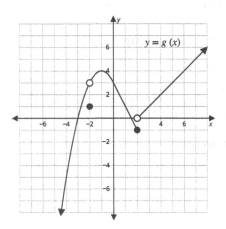

Solution

Example 3: Finding Limits Algebraically

a. Determine $\lim\limits_{x \to 2} \left(\dfrac{10\left(x^2 - 4\right)}{x - 2} \right)$ by using an algebraically equivalent expression.

Solution

b. Determine $\lim\limits_{x \to 5} \left(\dfrac{x^3 - 125}{\left(x - 5\right)^3} \right)$ algebraically.

Solution

c. Determine $\lim\limits_{x \to -3} \left(\dfrac{2x^2 + 5x - 3}{x + 3} \right)$ algebraically.

Solution

Indeterminate Form

A limit expression of the type $\lim\limits_{x \to a} \left(\dfrac{g\left(x\right)}{f\left(x\right)} \right)$ is called an **indeterminate form** of type $\dfrac{0}{0}$ if

$\lim\limits_{x \to a} g\left(x\right) = $ _____ and $\lim\limits_{x \to a} f\left(x\right) = $ _____ .

10.2 Exercises

1. Use the graph to find the indicated limits.

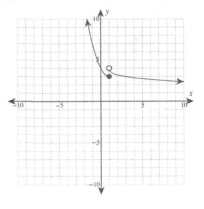

a. Find $\lim\limits_{x \to 1^-} f(x)$.

b. Find $\lim\limits_{x \to 1^+} f(x)$.

c. Find $\lim\limits_{x \to 1} f(x)$.

2. Use the graph to find the indicated limits.

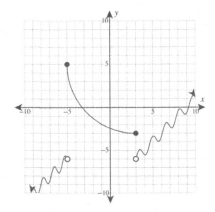

a. Find $\lim\limits_{x \to -5^-} f(x)$.

b. Find $\lim\limits_{x \to -5^+} f(x)$.

c. Find $\lim\limits_{x \to -5} f(x)$.

d. Find $\lim\limits_{x \to 3} f(x)$.

3. Use the graph to find the indicated limits.

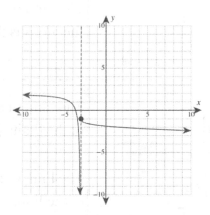

a. Find $\lim\limits_{x \to -3^-} f(x)$.

b. Find $\lim\limits_{x \to -3^+} f(x)$.

c. Find $\lim\limits_{x \to -3} f(x)$.

4. Find the limit algebraically by factoring the expression first.

$$\lim_{x \to 3} \left(\frac{3x^2 - 6x - 9}{x - 3} \right)$$

10.3 More about Limits

Limits as $x \to a$

Suppose that c is any constant, n is a positive integer, a is a real number, $\lim_{x \to a} f(x) = L_1$, and $\lim_{x \to a} g(x) = L_2$, where L_1 and L_2 are real numbers.

Property	Comments
1. $\lim_{x \to a} c =$ _____	The limit of a _____ is the constant itself.
2. $\lim_{x \to a} x =$ _____	
3. $\lim_{x \to a} x^n =$ _____	
4. $\lim_{x \to a} [f(x) \pm g(x)]$ $= \lim_{x \to a} f(x) \pm \lim_{x \to a} g(x) =$ _____	The limit of a sum (or difference) of two functions is the _____ (or _____) of the limits.
5. $\lim_{x \to a} [f(x) \cdot g(x)]$ $= \left[\lim_{x \to a} f(x) \right] \cdot \left[\lim_{x \to a} g(x) \right] =$ _____	The limit of a product is the _____ of the limits provided these limits exist.
6. $\lim_{x \to a} \left[\dfrac{f(x)}{g(x)} \right] = \dfrac{\lim_{x \to a} f(x)}{\lim_{x \to a} g(x)} =$ _____ where $L_2 \neq 0$	The limit of a quotient is the _____ of the limits as long as the limit of the denominator is not 0.
7. $\lim_{x \to a} [c \cdot f(x)] = c \cdot \lim_{x \to a} f(x) =$ _____	The limit of a constant times a function can be found by first finding the limit of the function and then _____ this limit by the constant.
8. $\lim_{x \to a} \sqrt[n]{f(x)} = \sqrt[n]{\lim_{x \to a} f(x)}$ $=$ _____ $=$ _____ where $(L_1)^{\frac{1}{n}}$ is defined.	The limit of the _____ of a function can be found by calculating the limit of the function first and then taking the nth root of the limit. (**Note:** $f(x)$ must be defined on intervals to the left of a and to the right of a.)
9. If p is a polynomial function, then $\lim_{x \to a} p(x) =$ _____ .	The limit of a polynomial can be found by substituting _____ for x. This result is particularly useful and follows directly from Properties 1 through 5.

Example 1: Properties of Limits

Use the properties of limits to find each of the following limits.

a. $\lim\limits_{x \to 4} 10$

Solution

b. $\lim\limits_{x \to 3} \left(x^2 + 5x + 7\right)$

Solution

c. $\lim\limits_{x \to 2} \sqrt[3]{\dfrac{x - 3}{x + 6}}$

Solution

Example 2: Indeterminate Form

If $f\left(x\right) = \dfrac{x^3 - 1}{x - 1}$, find $\lim\limits_{x \to 1} f\left(x\right)$.

Solution

Example 4: Piecewise Function

The function $f(x)$ is defined piecewise as follows:

$$f(x) = \begin{cases} 2x & \text{if } x \leq 3 \\ x+3 & \text{if } x > 3 \end{cases}.$$

Find $\lim\limits_{x \to 3} f(x)$ if it exists.

Solution

Rational Function

A rational function is a function of the form

$$f(x) = \frac{p(x)}{q(x)},$$

where $p(x)$ and $q(x)$ are _____.

Example 7: Properties of Limits

Find $\lim\limits_{x \to -\infty} \left(\dfrac{x^3 + x^2 - x + 1}{x^2 - 4} \right)$, if it exists.

Solution

Summary of Limits for Rational Functions as $x \to +\infty$ or $x \to -\infty$

Consider the function

$$f(x) = \frac{a_n x^n + a_{n-1} x^{n-1} + \ldots + a_0}{b_m x^m + b_{m-1} x^{m-1} + \ldots + b_0},$$

where $a_n \neq 0$ and $b_m \neq 0$.

Case 1: For $m = n$, $\lim\limits_{x \to +\infty} f(x) = $ _____ .

Case 2: For $m > n$, $\lim\limits_{x \to +\infty} f(x) = $ ____ .

Case 3: For $m < n$, $\lim\limits_{x \to +\infty} f(x) = $ _____ . (Or $-\infty$ depending on the signs of a_n and b_m.)

10.3 Exercises

1. Find the indicated limit, if it exists.

$$\lim_{x \to 2^+} \left(\frac{-5 + x}{10x - 2} \right)$$

2. Find the indicated limit, if it exists.

$$\lim_{x \to 0^-} \left(\frac{x}{-2x^3 + 11x} \right)$$

3. Find the indicated limit, if it exists.

$$\lim_{x \to 0^+} \left(\frac{14x^3 + 15x}{x} \right)$$

4. Find the indicated limit, if it exists.

$$\lim_{x \to +\infty} \left(\frac{-8x}{-14x^2 + 6} \right)$$

5. Find the indicated limit, if it exists.

$$\lim_{x \to +\infty} \left(\frac{x^3 + 343}{x^2 - 7x + 49} \right)$$

6. Find the indicated limit, if it exists.

$$\lim_{x \to 4} \left(\frac{x^2 - 7x + 12}{x^2 - 16} \right)$$

7. Find the indicated limit, if it exists.

$$\lim_{x \to 0^-} \left(-17 + \frac{10}{x^4} \right)$$

8. Find the indicated limit, if it exists.

$$\lim_{x \to -6^+} \left(-9x - \frac{18}{x + 6} \right)$$

9. Find the indicated limit, if it exists.

$$\lim_{x \to -\infty} \left(14 + \frac{9}{x^2} \right)$$

10. Find the indicated limit, if it exists.

$$\lim_{x \to -\infty} \left(\frac{10x^2 + 4x + 12}{12x^3 + 14x^2 - 6x + 19} \right)$$

10.4 Continuity

Continuity

A function $y = f(x)$ is _____ at $x = a$ if all of the following conditions are true:

1. $f(a)$ exists. ($f(a)$ is a _____ .)

2. $\lim\limits_{x \to a} f(x)$ exists. (The limit is a _____ .)

3. $\lim\limits_{x \to a} f(x) = f(a)$. (The limit is equal to the _____ of the function at $x = a$.)

If a function fails to satisfy any one of the three conditions of the definition, then it is _____ continuous at $x = a$ and is said to be _____ at $x = a$.

Example 8: Continuity

Use the definition of continuity to determine whether the function $f(x) = \dfrac{x^2 - x - 6}{x - 3}$ is continuous at

a. $x = 2$, and

b. $x = 3$.

Solution

Example 9: Graphing a Discontinuous Function

a. Use the definition of continuity to determine whether the function $g(x) = \dfrac{|x|}{x}$ is continuous at $x = 0$.

Solution

b. Draw the graph of $g(x)$.

Solution

Example 10: Making a Function Continuous

Find a value for k so that the function

$$f(x) = \begin{cases} x^2 & \text{if } x < 2 \\ -3x + k & \text{if } x \geq 2 \end{cases}$$

will be continuous at $x = 2$.

Solution

10.4 Exercises

1. Use the graph of $y = f(x)$ to answer the question regarding the function.

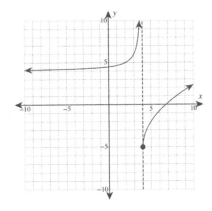

a. Find $\lim\limits_{x \to 4^-} f(x)$.

b. Find $\lim\limits_{x \to 4^+} f(x)$.

c. Find $f(4)$.

d. Is $f(x)$ continuous at $x = 4$?

2. Use the graph of $y = f(x)$ to answer the question regarding the function.

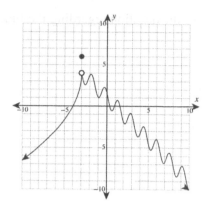

a. Find $\lim\limits_{x \to -3^-} f(x)$.

b. Find $\lim\limits_{x \to -3^+} f(x)$.

c. Find $f(-3)$.

d. Is $f(x)$ continuous at $x = -3$?

3. Find the value for k so that the function will be continuous at $x = 2$.

$$f(x) = \begin{cases} -2x & \text{if } x < 2 \\ -x^2 + 20x - 100 + k & \text{if } x \geq 2 \end{cases}$$

4. Consider the following function. Identify the number of points of discontinuity for $f(x)$. Then find the x-values of any points of discontinuity and classify each discontinuous point as a non-removable discontinuity, removable discontinuity, or jump discontinuity.

$$f(x) = \frac{x - 10}{x - 5}$$

10.5 Average Rate of Change

Average Rate of Change

If $y = f(x)$ then the ratio

$$\frac{\Delta y}{\Delta x} = \underline{\hspace{4cm}} = \frac{f(x_2) - f(x_1)}{x_2 - x_1}$$

is called the average _____ **of y with respect to x** as x changes from x_1 to x_2.

Example 1: Finding the Average Rate of Change

For the function $y = 2x + 1$, find the average rate of change of y with respect to x, $\dfrac{\Delta y}{\Delta x}$, as x changes from 0 to 3.

Solution

Example 2: Average Rate of Change

Let $f(x) = \dfrac{2}{x}$. Find and simplify the expression that represents the average rate of change of f between x and $x + h$.

Solution

Example 3: A Falling Object

A ball is dropped from an airplane. The distance d that it falls in t seconds is given by the function $d = 16t^2$. The table shows the distance the ball falls and the change in distance Δd for 1-second time intervals during the first 5 seconds of its descent.

a. How far does the ball fall during the first 3 seconds?

b. What is the average velocity of the ball during the first 3 seconds? (Velocity is the rate of change of distance with respect to time.)

c. How fast is the ball traveling when $t = 3$?

t (sec)	$d = 16t^2$	Δd (ft)
0	0	—
1	16	$d_1 - d_0 = 16 - 0 = 16$
2	64	$d_2 - d_1 = 64 - 16 = 48$
3	144	$d_3 - d_2 = 144 - 64 = 80$
4	256	$d_4 - d_3 = 256 - 144 = 112$
5	400	$d_5 - d_4 = 400 - 256 = 144$

Solution

10.5 Exercises

1. Consider the graph of $f(x)$.

What is the average rate of change of $f(x)$ from $x_1 = 4$ to $x_2 = 7$? Please write your answer as an integer or simplified fraction.

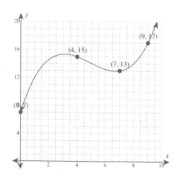

2. Consider the graph of $f(x)$.

What is the average rate of change of $f(x)$ from $x_1 = 1$ to $x_2 = 8$? Please write your answer as an integer or simplified fraction.

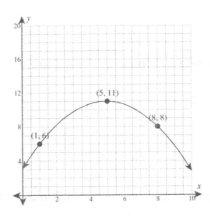

3. Consider the graph of $f(x)$.

What is the average rate of change of $f(x)$ from $x_1 = 4$ to $x_2 = 9$? Please write your answer as an integer or simplified fraction.

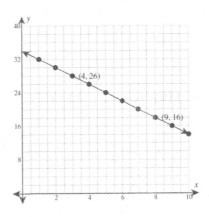

4. What is the average rate of change of $f(x)$ from $x_1 = 8$ to $x_2 = 10$? Please write your answer as an integer or simplified fraction.

$f(x) = 3x - 6$

5. What is the average rate of change of $f(x)$ from $x_1 = -5.7$ to $x_2 = -1.6$? Please write your answer rounded to the nearest hundredth.

$$f(x) = -7x - 1$$

6. What is the average rate of change of $f(x)$ from $x_1 = -4.1$ to $x_2 = -1.1$? Please write your answer rounded to the nearest hundredth.

$$f(x) = -8x^2 + 4x + 9$$

7. What is the average rate of change of $f(x)$ from $x_1 = -3.7$ to $x_2 = 6.2$? Please write your answer rounded to the nearest hundredth.

$$f(x) = 2.1x^2 + 0.5x - 1.1$$

8. What is the average rate of change of $f(x)$ from $x_1 = 0.3$ to $x_2 = 5.4$? Please write your answer rounded to the nearest hundredth.

$$f(x) = 5x^3 - 7x^2 + 6x + 2$$

9. What is the average rate of change of $f(x)$ from $x_1 = 4$ to $x_2 = 6.9$? Please write your answer rounded to the nearest hundredth.

$$f(x) = \frac{4}{8x - 7}$$

10. What is the average rate of change of $f(x)$ from $x_1 = 1$ to $x_2 = 4$? Please write your answer rounded to the nearest hundredth.

$$f(x) = \sqrt{4x + 2}$$

10.6 Instantaneous Rate of Change

Slope of a Point on a Curve

The _____ of a curve $f(x)$ at the point (a, b) is exactly the slope of the _____ to the curve at that point. This is also called the _____ **rate of change.**

Difference Quotient

If $(x, f(x))$ is any fixed point on a curve and $(x + h, f(x + h))$ is another point on the curve, then the _____ is the slope of the secant line through these two points:

$$m_{\text{sec}} = \frac{f(x + h) - f(x)}{(x + h) - x} = \frac{f(x + h) - f(x)}{h}.$$

Example 5: Manufacturing

Let $C(x)$ denote the total cost of manufacturing x units of a certain style of metal office bookcase.

a. Given $C(300) = 31{,}000$ and $C'(300) = 90$, sketch a tangent line to $C(x)$ at the point $(300, 31{,}000)$.

b. Estimate the cost of manufacturing 303 bookcases.

Solution

Example 6: Average Rate of Change

Consider the function $f(x) = \frac{1}{3}x^3$. Find the average rate of change of f as x changes from $x_1 = 0$ to $x_2 = 3$.

Solution

Example 7: Instantaneous Velocity

An arrow is shot into the air and its height in feet after t seconds is given by the function $f(t) = -16t^2 + 80t$. The graph of the curve $y = f(t)$ is the parabola shown on the right.

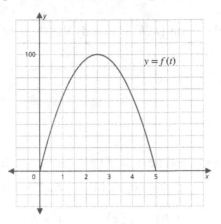

a. Find the height of the arrow when $t = 3$.

b. Find the velocity of the arrow when $t = 3$.

c. Find the slope of the tangent line to the curve at $t = 3$.

d. Find the time it takes the arrow to reach its peak.

Solution

10.6 Exercises

1. $f(x)$ is the total cost of manufacturing x tables. Which of the following does the slope $f'(x)$ represent?

- ○ Total revenue from the sale of x tables.

- ○ Additional cost for increasing production by one table when x tables are being produced.

- ○ Rate of change in the cost of increasing production by one when x tables are being produced.

- ○ Average rate of change in the cost of manufacturing x tables.

2. Find the slope of $f(x) = -4x - 3$ at $(1, -7)$.

3. Consider the function $f(x) = x^3 - 10x^2 + 9x - 2$.

a. Interpret the meaning of $f(7) = -86$.

b. Interpret the meaning of $f'(7) = 16$.

4. Suppose $f(x)$ is the total number of children that have a disease, and x is the number of days since the first case.

a. Interpret the meaning of $f(36) = 900$.

b. Interpret the meaning of $f'(36) = 19$.

5. The total number of births in an average woman's lifetime in India can be approximated by the function $f(x) = -0.068x + 5.22$, where x is the number of years after 1970 and corresponds to the woman's year of birth.

a. What is the value of $f(19)$ and what does it represent?

b. What is the value of $f'(19)$ and what does it represent?

6. Find the slope of $f(x) = 4\sqrt{x} - 3x$ at $(7, -10.42)$. Use a graphing utility and round to the nearest hundredth.

7. Find the slope of $f(x) = \ln(3x)$ at $(8, 3.18)$. Use a graphing utility and round to the nearest hundredth.

8. Find the slope of $f(x) = \dfrac{-2x - 1}{8\sqrt{x}}$ at $(9, -0.79)$. Use a graphing utility and round to the nearest hundredth.

9. Find the slope of $f(x) = \ln(7x^2 + 8)$ at $(5, 5.21)$. Use a graphing utility and round to the nearest hundredth.

10.7 Definition of the Derivative and the Power Rule

Derivative

For any given function $y = f(x)$, the _____ of f at x is defined to be

$$f'(x) = \lim_{h \to 0} \left(\frac{f(x+h) - f(x)}{h} \right),$$

provided this limit exists. (**Note:** $\Delta x = (x + h) - x = h$.)

The Three-Step Method for Finding a Derivative

1. Form the ratio $\dfrac{f(x+h) - f(x)}{h}$, called the _____.

2. Simplify the difference quotient algebraically.

3. Calculate $f'(x) = $ _____ $\left(\dfrac{f(x+h) - f(x)}{h} \right)$, if it exists.

Example 1: Finding a Derivative

Use the Three-Step Method to find $f'(x)$ for $f(x) = x^3 - 5x$.

Solution

Conditions of a Non-Differentiable Function

A function $f(x)$ is _____ at $x = a$ if any of the following is true:

1. $f(x)$ is _____ at $x = a$.

2. The graph of $f(x)$ has a _____ at $x = a$.

3. The tangent line at $x = a$ is a _____ line.

Continuity at $x = a$

The function f is _____ at $x = a$ if and only if

$$\lim_{x \to a^-} (f(x)) = f(a) = \lim_{x \to a^+} (f(x)).$$

Continuous Function on an Interval

A function f is **continuous on an** _____ (x_1, x_2) provided the function is

_____ at $x = a$ for every number a with $x_1 < a < x_2$.

The Power Rule

For the function $f(x) = x^r$, where r is any real number, then

_____ .

Basic Rule of the Derivative of a Linear Function

For $y = mx + b$, $y' =$ _____ .

Two Special Cases of the Basic Rule

1. If $f(x) = c$, then $f'(x) =$ _____ .   Has slope 0 (constant functions are horizontal lines).

2. If $f(x) = x$, then $f'(x) =$ _____ .   Has slope 1.

Polynomial Functions

A polynomial function is a function of the form

$$f(x) = a_0 + a_1 x + a_2 x^2 + \ldots + a_n x^n,$$

where $a_0, a_1, a_2, \ldots, a_n$ are real numbers and the exponents are positive integers.

Also, for any real number a, $\lim_{x \to a} f(x) = f(a)$, and therefore, $f(x)$ is _____ everywhere.

Power Rule (General Case)

For the function $f(x) = c \cdot x^r$ where c and r are any real numbers,

_____ .

Example 4: Finding the Derivatives of Functions

Find the derivative of each of the following functions.

a. $f(x) = 5$

Solution

b. $f(x) = 5x^3$

Solution

c. $f(x) = x^{\frac{1}{2}}$

Solution

Example 5: Rewriting a Fraction to Find the Derivative

Find the derivative of $g(x)$ given that $g(x) = \dfrac{1}{x}$.

Solution

Example 6: Using Derivative Notation

Find $D_x\left(h\left(x\right)\right)$ where $h\left(x\right) = \dfrac{18}{x^{\frac{2}{3}}}$.

Solution

10.7 Exercises

1. Find the derivative for the following function.

$y = -5x$

2. Find the derivative for the following function.

$y = 4x - 5$

3. Find the derivative for the following function.

$f\left(x\right) = 7x^4$

4. Find the derivative for the following function.

$f\left(x\right) = \dfrac{6}{x^5}$

5. Find the derivative for the following function.

$f\left(x\right) = 8\sqrt[3]{x}$

6. Find the derivative for the following function.

$$y = \frac{-3}{\sqrt[5]{x}}$$

7. Find the derivative for the following function.

$$y = 6x^{0.2}$$

8. Find the derivative for the following function.

$$f(x) = -2x^{6.4}$$

9. Find the derivative for the following function.

$$f(x) = -8x^{\frac{2}{7}}$$

10. Find the derivative for the following function.

$$g(x) = x^{\frac{4}{3}}$$

10.8 Techniques for Finding Derivatives

Constant Times a Function Rule

If $f(x)$ is a differentiable function, c is a real constant, and $y = c \cdot f(x)$, then

$$\frac{dy}{dx} = c \cdot f'(x).$$

In words, the derivative of a constant times a function is the _____ times the

_____ of the function. (The power rule is a special case of this rule.)

The Sum and Difference Rule

If $f(x)$ and $g(x)$ are differentiable functions and $y = f(x) \pm g(x)$, then

$$\frac{dy}{dx} = f'(x) \pm g'(x).$$

In words, the derivative of a sum (or difference) of two functions is the _____

_____ of their derivatives.

Example 7: Using the Rules

Find the derivative of each of the following functions.

a. $y = 5x^2$

b. $y = -4x$

c. $y = 5x^2 - 4x$

d. $y = x^3 + 8\sqrt{x} + \dfrac{2}{x}$

Solution

Example 8: Algebraic Manipulations

a. Find $f'(u)$ if $f(u) = u^{\frac{1}{2}}\left(u^2 + 2u\right)$.

Solution

b. If $F(v) = \dfrac{v^2 + 1}{\sqrt{v}}$, find $F'(v)$.

Solution

Example 9: Tangent Lines

For the function $f(x) = 8x - x^2$:

a. Find the slope of the tangent lines at $x = 2$, $x = 4$, and $x = 5$.

b. Find the equation of the tangent lines at $x = 2$, $x = 4$, and $x = 5$.

Solution

a.

b.

$$\rule{7cm}{0.8pt}$$

Example 10: Velocity

Suppose that a sailboat is observed, over a period of 5 minutes, to travel a distance from a starting point according to the function $s\,(t) = t^3 + 60t$, where t is time in minutes and s is the distance traveled in meters.

a. How fast is the boat moving at the starting point?

b. How fast is the boat moving at the end of 3 minutes?

Solution

Example 11: Identifying Derivatives in Real-Life

Determine which of the following represent (or might represent) the value of a derivative at a point, and if so, give a suitable description of a function f.

a. 30 miles per hour.

b. 6 fish per day.

c. 23 ears of corn.

d. 19 neutrons per millisecond.

Solution

10.8 Exercises

1. Find the derivative for

$$f(x) = -4x^5 + 2x^3$$

2. Find the derivative for

$$g(x) = 9x^4 - 4.8 + 9x^3$$

3. Find the derivative for

$$g(x) = \frac{1}{5}x^4 + \frac{3}{2} + 5x^5 + 8x^2$$

4. Find the derivative for

$$f(x) = -5x^{\frac{2}{3}} + 3x^{\frac{1}{2}} + 3$$

5. Use algebraic techniques to rewrite $f(x) = (4x + 5)(2x + 4)$ as a sum or difference; then find $f'(x)$.

6. Use algebraic techniques to rewrite $y = x^{\frac{1}{2}}\left(8x^{\frac{5}{2}} - 6x^{\frac{7}{2}} - 3\right)$ as a sum or difference; then find y'.

7. A town's population t years from now can be estimated from the formula $P(t) = 8900 + 500t - 92\sqrt{t}$.

Find the rate at which the town is growing after 4 years.

8. A tenant dropped a spoon from the top floor of her apartment building and it fell according to the formula $S(t) = -16t^2 + 2t^{0.3}$, where t is the time in seconds and $S(t)$ is the distance in feet from the top of the building.

    a. If the spoon hit the ground in exactly 2.3 seconds, how high is the building? Round your answer to 2 decimal places.

    b. What was the average speed of the fall? Use the fact that the spoon hit the ground in exactly 2.3 seconds. Round your answer to 2 decimal places.

    c. What was the instantaneous velocity at $t = 2$ seconds? Round your answer to 2 decimal places.

10.9 Applications: Marginal Analysis

Marginal Cost

Marginal cost is the rate of change of _____ per unit change in _____ .
This concept is illustrated geometrically in Figure 2.4.1.

Example 1: Marginal Cost

A speciality item manufacturer determines that the cost of producing x ballpoint pens is $C(x) = 500 + 3x$ in dollars.

    a. Find $C(101) - C(100)$, the cost of producing the 101st pen.

    b. Find $C'(100)$, the marginal cost at $x = 100$ pens.

Solution

Marginal Revenue

Marginal revenue is the rate of change of the _____ per unit change in _____
when the level of sales is x items.

Example 2: Marginal Revenue

Suppose that a manufacturer has determined that the price of certain custom-made tables he produces can be determined by the demand function $p = D(x) = 117 - \frac{x}{4}$, where x is the number of tables produced and sold.

a. Find the revenue function.

b. Determine the marginal revenue when $x = 16$ tables.

Solution

Marginal Profit

Marginal profit is the rate of change of the _____ per unit change in _____ when x items are produced and sold.

Marginal Average Cost

Marginal average cost $\overline{C}'(x)$ is the _____ per unit change in _____. This is an approximation of the _____ _____ when one more item is produced.

Example 4: Marginal Average Cost

Suppose $C(x) = 0.02x^2 + 100x + 2000$ represents the cost in dollars of producing x radios.

   a. Find the average cost function.

   b. Find the average cost of producing 1000 radios.

   c. Find the marginal average cost if 1000 radios are produced.

Solution

Example 5: Marginal Propensity

If the marginal propensity to save is $S'(x) = 0.02x$, what is the marginal propensity to consume?

Solution

10.9 Exercises

   1. The cost of producing x grapefruits is given by $C(x) = 830 + 2x$. Determine the average cost function.

   2. A hotdog vendor has initial start up costs of $7300 and an item charge of $0.50 per combo meal. What is the cost function?

   3. A vendor charges $42 for a hat. What is the demand function?

4. A vendor charges \$15 for a souvenir. What is the revenue function?

5. A supply company charges for 90-liter drums using the demand function $p(x) = 20.9 - 1.6x$. What is the revenue function?

6. A supplier of souvenir caps charges street vendors for each order based on a set up fee of \$34 and an item charge of \$1.8 per cap. What is the cost function for the vendor?

7. The weekly cost of producing x magazines is given by the function $C(x) = 2500 + 21x + 0.4x^2$. Find the marginal average cost function.

8. The cost of producing x units of a commodity is given by $C(x) = 80 + 18x - 0.8x^2$. Find the marginal cost function.

9. A company can sell x units of a product when the price is $p = D(x) = 3.00 - 0.001x$ dollars. The total production costs are $C(x) = 0.004x^2 + 0.61x + 280$ dollars. Find the marginal profit function.

10. A sales representative for a company that produces electric mixers can sell x units of their deluxe model if the price is $p = D(x) = 86.9 - 0.02x$ dollars. The total cost for these electric mixers is given by $C(x) = 0.08x^2 + 4.2x + 5800$ dollars. Determine the marginal profit for 87 electric mixers.

11.1 The Product and Quotient Rules

Product Rule

If $f(x)$ and $g(x)$ are differentiable functions and $y = f(x) \cdot g(x)$, then

$$\frac{dy}{dx} = f(x) \cdot g'(x) + g(x) \cdot f'(x).$$

In words, the derivative of the product of two functions is equal to the _____ times

the _____ plus the _____ times the

_____ .

Other Forms of the Product Rule

If $f(x)$ and $g(x)$ are differentiable functions and $y = f(x) \cdot g(x)$, then

1. $\dfrac{dy}{dx} = $ _____ $+ g(x) \cdot \dfrac{d}{dx}[f(x)],$

2. $\dfrac{d}{dx}[f(x) \cdot g(x)] = f(x) \cdot$ _____ $+ g(x) \cdot$ _____ ,

3. $y' = $ _____ $\cdot g'(x) + $ _____ $\cdot f'(x),$

4. $d_x[f(x) \cdot g(x)] = f(x) \cdot$ _____ $+ g(x) \cdot$ _____ , and

5. $y' = $ _____ $\cdot g' + $ _____ $\cdot f'.$

Example 1: Using the Product Rule

Use the Product Rule to find $\dfrac{dy}{dx}$ given $y = (x^2 + 3x - 1)(x^3 - 8x)$.

Solution

Example 2: Using the Product Rule

Use the Product Rule to find the derivative of the function $f\left(x\right) = \left(\sqrt{x} + x^{-1}\right)\left(x^2 + 1\right)$.

Solution

Quotient Rule

If $f\left(x\right)$ and $g\left(x\right)$ are differentiable functions and $y = \dfrac{f\left(x\right)}{g\left(x\right)}$, then

$$\frac{dy}{dx} = \frac{g\left(x\right) \cdot f'\left(x\right) - f\left(x\right) \cdot g'\left(x\right)}{\left(g\left(x\right)\right)^2}.$$

In words, the derivative of a quotient is the _____ times the _____ _____ minus the _____ times the _____ , all divided by the _____ of the denominator.

Example 3: Using the Quotient Rule

Use the Quotient Rule to find the derivative of the function $y = \dfrac{x^2}{5x + 1}$.

Solution

Example 5: Growth of Bacteria

Several years of study have shown that t hours after a bacterium is introduced to a particular culture, the number of bacteria is given by $N\left(t\right) = \dfrac{t^2 + t}{\sqrt{t} + 1}$. Find the rate of growth of the bacteria after 4 hours.

Solution

Caution

The algebraic form of the derivative of a function depends on which _____

_____ is applied, and there may be several correct forms.

For example, if $f\left(x\right) = \dfrac{x^3 + 3x + 1}{\left(2x + 5\right)}$, then we will find (in this section) that the following two forms for $f'\left(x\right)$ are correct.

1. $f'\left(x\right) = \dfrac{\left(2x + 5\right)\left(3x^2 + 3\right) \underline{\hspace{1cm}} \left(x^3 + 3x + 1\right)\left(2\right)}{\left(2x + 5\right)^2}$

2. $f'\left(x\right) = \left(x^3 + 3x + 1\right)\left(-1\right)\left(2x + 5\right)^{-2}\left(2\right) \underline{\hspace{1cm}} \left(2x + 5\right)^{-1}\left(3x^2 + 3\right)$

   Both of these forms can be simplified to the form

3. $f'\left(x\right) = \dfrac{4x^3 + 15x^2 + 13}{\left(2x + 5\right)^2}$.

If the rules of differentiation have been followed, then an answer may be correct even though it does not "look like" the answer given in the Answers section at the end of the text. Be sure to check with your instructor to see how much algebraic simplification is expected and whether or not one form is preferred over another.

11.1 Exercises

1. Use the Product Rule or Quotient Rule to find the derivative.

$$f(x) = \frac{-9x^4 + 27x^3 - 21x + 63}{-3x + 9}$$

2. Use the Product Rule or Quotient Rule to find the derivative.

$$f(x) = \left(\frac{5}{x^{\frac{1}{2}}} + \frac{1}{x^2} \right) (-2x^3 + 1)$$

3. Given $f(4) = 8$, $f'(4) = -18$, $g(4) = -16$, and $g'(4) = -8$, find the value of $h'(4)$ based on the function below.

$$h(x) = f(x) \cdot g(x)$$

4. Given $g(3) = 16$ and $g'(3) = -4$, find the value of $h'(3)$ based on the function below.

$$h(x) = g(x) \cdot (-2x + 10)$$

5. Given $f(4) = 3$, $f'(4) = -6$, $g(4) = 15$, and $g'(4) = -9$, find the value of $h'(4)$ based on the function below.

$$h(x) = \frac{g(x) + 4x - 4}{f(x)}$$

6. It is estimated that t years from now the population of a city will number $P(t) = (0.9t - 3)(0.7t + 9) + 85$ thousand people. How fast will the population (in thousands) be growing in 9 years? Round your answer to two decimal places.

7. Use the Product Rule or Quotient Rule to find the derivative.

$$f\left(x\right) = \frac{-x^3 + 1}{8x^2 + 4}$$

8. Use the Product Rule or Quotient Rule to find the derivative.

$$f\left(x\right) = x^2\left(-7x^5 - 7\right)$$

9. Find the equation of the tangent line to the graph of $f\left(x\right)$ at the (x, y)-coordinate indicated below.

$$f\left(x\right) = \left(-x^2 + 4\right)\left(-3x^2 + 4x - 4\right); \left(-1, -33\right)$$

10. Given $f\left(x\right) = \left(2x - 3\right)\left(-3x - 2\right)$, find the (x, y)-coordinate on the graph where the slope of the tangent line is -5.

11.2 The Chain Rule and the General Power Rule

The Chain Rule

If $y = f(u)$ and $u = g(x)$, then

$$\underline{\hspace{6cm}},$$

provided that $\dfrac{dy}{du}$ and $\dfrac{du}{dx}$ both exist.

Example 1: Using the Chain Rule

Suppose that $y = u + \sqrt{u}$ and $u = x^3 + 17$. Use the Chain Rule to find $\dfrac{dy}{dx}$. Then evaluate $\dfrac{dy}{dx}$ at $x = 2$. This evaluation is denoted more briefly by $\dfrac{dy}{dx}\bigg|_{x=2}$.

Solution

The Chain Rule (Second Form)

If $y = f(g(x))$, then

$$\underline{\hspace{6cm}}$$

provided that $f'(g(x))$ and $g'(x)$ both exist.

Example 3: Second Form of the Chain Rule

Use the second form of the Chain Rule to find $\dfrac{dy}{dx}$ if $y = \left(x^2 + 3x - 7\right)^{\frac{5}{2}}$.

Solution

The General Power Rule

If $y = [g(x)]^n$, then

_____

provided that $g'(x)$ exists.

Example 5: The General Power Rule

Using the General Power Rule, find the derivative of each of the following functions.

a. $y = (x^2 + 8x)^{10}$

Solution

Example 6: The General Power Rule

Use the General Power Rule to find $f'(u)$ if $f(u) = \dfrac{1}{\sqrt{7-u}}$.

Solution

Example 8: Using Multiple Rules

Find $d_x \left[\sqrt[3]{\dfrac{2x-7}{3x+4}} \right]$.

Solution

11.2 Exercises

1. Find the derivative for the given function. Write your answer using positive and negative exponents and fractional exponents instead of radicals.

$$f\left(x\right) = \left(-x^3 + 4x + 5\right)\left(5x^4 - 2x + 3\right)^3$$

2. Find the derivative for the given function. Write your answer using positive and negative exponents and fractional exponents instead of radicals.

$$h\left(x\right) = \frac{\left(3x^2 + x + 6\right)^{\frac{2}{3}}}{5x^{-3} - 4x + 8}$$

3. Find the derivative for the given function. Write your answer using positive and negative exponents and fractional exponents instead of radicals.

$$f\left(x\right) = \left(\frac{7x^3 - 9x + 5}{x^3 - 5x + 10}\right)^{-3}$$

4. Given $f\left(8\right) = 9$ and $f'\left(8\right) = 2$ find the value of $h'\left(8\right)$ based on the function below.

$$h\left(x\right) = \left(f\left(x\right)\right)^3$$

5. Find the derivative for the given function. Write your answer using positive and negative exponents instead of fractions and use fractional exponents instead of radicals.

$$h\left(x\right) = \left(\left(-2x^3 + 9\right)\left(7x^2 + 7\right)\right)^3$$

6. Find the derivative for the given function. Write your answer using positive and negative exponents and fractional exponents instead of radicals.

$$y = \sqrt[4]{8x^3 + 7}$$

7. Find the derivative for the given function. Write your answer using positive and negative exponents and fractional exponents instead of radicals.

$$g(t) = \frac{7}{\sqrt{t^3 - 4}}$$

8. Find the derivative for the given function. Write your answer using positive and negative exponents and fractional exponents instead of radicals.

$$h(x) = \frac{\sqrt[3]{3x + 4}}{x^2}$$

9. After a sewage spill, the level of pollution in San Remo Bay is estimated by $f(t) - \dfrac{200t^2}{\sqrt{t^2 + 16}}$, where t is the time in days since the spill occurred. How fast is the level changing after 6 days? Round to the nearest whole number.

10. Determine the equation of the tangent line for $f(x) = \left(2x^2 + 3x - 6\right)^2$ at the x-value given below.

$$x = -2$$

11.3 Implicit Differentiation and Related Rates

Example 2: Implicit Differentiation and Related Rates

If $xy + x^2y^3 - 1 = 0$, find $\dfrac{dy}{dx}$.

Solution

Example 3: Implicit Differentiation

Find $\dfrac{dy}{dx}$ if $\sqrt{x} - \sqrt{y} = 16$.

Solution

Example 5: Airplane Holding Pattern

An airplane is flying in a circular "holding" pattern with a radius of 2 miles. The path of the plane can be described by the equation $x^2 + y^2 = 4$. When the plane is at the point $x = \sqrt{3}$ miles and $y = 1$ mile, as shown in the figure, it is traveling north at 300 mph. How fast is it traveling west?

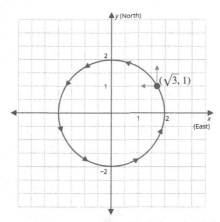

Solution

Example 6: Moving Cars

Two cars, A and B, are traveling toward the same intersection, and neither driver plans to stop. Car A is 0.5 miles from the intersection moving east at 40 mph, and Car B is 0.75 miles from the intersection moving south at 60 mph. How fast is the distance between the two cars changing?

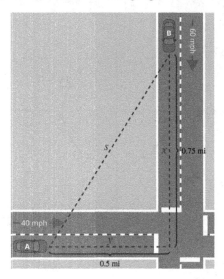

Solution

Cobb-Douglas Production Formula

$$P = Cx^a y^{1-a}$$

where

$P = $ _____ ,

$x = $ _____ , and

$y = $ _____ .

C and a are constants, $0 < a < 1$.

Example 7: Production

Suppose that a firm's level of production is given by $P = 20x^{\frac{1}{4}}y^{\frac{3}{4}}$, where x represents the units of labor and y represents the units of capital. Currently, the company is using 16 units of labor and 81 units of capital. If labor is increasing by 4 units per month, what must the change in units of capital per month be to maintain the current level of production?

Solution

11.3 Exercises

1. Find $\dfrac{dy}{dx}$ for the equation below.

$2x^2 - 6y^3 = 8$

2. Find $\dfrac{dy}{dx}$ for the equation below.

$7\sqrt[3]{x} + \sqrt{y} = 4$

3. Find $\dfrac{dy}{dx}$ for the equation below.

$6x^2y^3 = 4$

4. Find $\dfrac{dy}{dx}$ for the equation below.

$\dfrac{2}{x^3} + \dfrac{x}{y} = 6x^4$

5. Consider the following equation.

$x^2 - y^2 = 3$

    a. Use implicit differentiation to find $\dfrac{dy}{dx}$.

    b. Find the slope of the tangent line at $(-2, -1)$.

6. Consider the following equation.

$3x^4 - 3xy + 2y^3 = 66$

    a. Use implicit differentiation to find $\dfrac{dy}{dx}$.

    b. Find the slope of the tangent line at $(-1, 3)$.

7. The level of production of a company is given by $31x^{\frac{2}{3}}y^{\frac{1}{3}}$, where x is the units of labor and y is the units of capital. The company is currently utilizing 8 units of labor and 27 units of capital. If labor is increased by 3 units per week, what will be the change in units of capital per week to maintain the current level of production?

8. A projectile traveling horizontally at a height of 1500 feet crosses directly over a woman. The speed of the projectile is 480 ft / sec. How fast is the distance between the projectile and the woman changing after 15 seconds? Round to the nearest integer.

11.4 Increasing and Decreasing Intervals

Increasing and Decreasing Intervals of a Function

Suppose that a function f is defined on the interval (a, b).

1. If $x_1 < x_2$ implies $f(x_1) < f(x_2)$ for every x_1 and x_2 in (a, b), then $f(x)$ is _____ on (a, b).

2. If $x_1 < x_2$ implies $f(x_1) > f(x_2)$ for every x_1 and x_2 in (a, b), then $f(x)$ is _____ on (a, b).

Example 1: Increasing and Decreasing Intervals

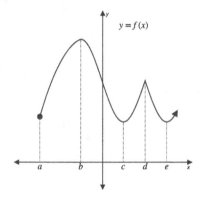

The graph of $y = f(x)$ is given. Find the intervals on which **a.** f is increasing, and **b.** f is decreasing.

Solution

Theorem

Suppose that $f(x)$ is differentiable on the interval (a, b).

If $f'(x)$ is positive ($f'(x)$ _____ 0) for all x in (a, b), then f is _____ on (a, b).

If $f'(x)$ is negative ($f'(x)$ _____ 0) for all x in (a, b), then f is _____ on (a, b).

Note: The values of x for which a function is increasing (or decreasing) are always presented as _____ intervals. These intervals are part (or all) of the domain of the function.

Example 4: Graphing a Function

For the function $f\left(x\right) = \dfrac{1}{x}$,

   a. Find all values of x where the derivative is 0.

   b. Determine where the function is increasing and where it is decreasing.

   c. Sketch its graph.

Solution

Example 5: Graphing $f'(x)$

The graph for $y = f\left(x\right)$ is given below. Mark the portions which show where $f'\left(x\right)$ is negative, positive, or zero. Draw a possible $f'\left(x\right)$ by estimating the absolute values of the slopes on $f\left(x\right)$.

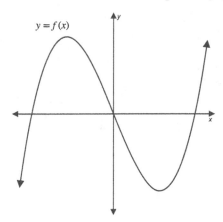

$y = f(x)$

Solution

11.4 Exercises

1. Consider the graph and determine the open intervals on which the function is increasing and on which the function is decreasing. Write ∅ to indicate the interval is empty.

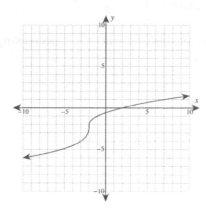

2. Consider the graph and determine the open intervals on which the function is increasing and on which the function is decreasing. Write ∅ to indicate the interval is empty.

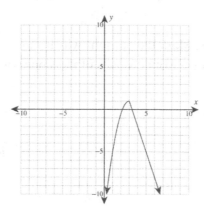

3. Consider the graph and determine the open intervals on which the function is increasing and on which the function is decreasing. Write ∅ to indicate the interval is empty.

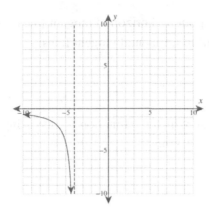

4. Consider the graph and determine the open intervals on which the function is increasing and on which the function is decreasing. Write ∅ to indicate the interval is empty.

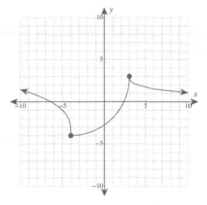

5. Consider the function.

$$f(x) = x^3 - 6x^2 + 12x - 10$$

    a. Find all values of x that correspond to horizontal tangent lines. Write "None" if the function does not have any values of x that correspond to horizontal tangent lines.

    b. Determine the open intervals on which the function is increasing and on which the function is decreasing. Write $\varnothing$ to indicate the interval is empty.

6. Consider the function.

$$f(x) = -4x^3 + 12x^2 + 180x$$

    a. Find all values of x that correspond to horizontal tangent lines. Write "None" if the function does not have any values of x that correspond to horizontal tangent lines.

    b. Determine the open intervals on which the function is increasing and on which the function is decreasing. Write $\varnothing$ to indicate the interval is empty.

7. Determine the open intervals on which the function is increasing and on which the function is decreasing. Write $\varnothing$ to indicate the interval is empty.

$$f(x) = \frac{x + 8}{x - 4}$$

8. Determine the open intervals on which the function is increasing and on which the function is decreasing. Write $\varnothing$ to indicate the interval is empty.

$$f(x) = \frac{-x^2 - 100}{x}$$

11.5 Critical Points and the First Derivative Test

Local Maximum and Local Minimum

Let f be a function defined at $x = c$.

1. $f(c)$ is called a local _____ (or relative _____) if there exists some interval (a, b) that contains c such that

$$f(x) \leq f(c)$$

for all x in (a, b). (We say a local maximum occurs at $x = c$.)

2. $f(c)$ is called a local _____ (or relative _____) if there exists some interval (a, b) that contains c such that

$$f(x) \underline{} f(c)$$

for all x in (a, b). (We say a local minimum occurs at $x = c$.)

Note: We also say that a local extremum occurs at the point $(c, f(c))$. or that the point $(c, f(c))$. is a local extremum, with the understanding that the y-value, $f(c)$, is actually the local extremum.

Theorem of Local Extrema

If a function f is continuous on the interval (a, b) and c is in (a, b) and $f(c)$ is either a local maximum or a local minimum, then

1. $f'(c) = \underline{}$, or

2. $f'(c)$ _____ .

Critical Values

_____ of x are those values c in the domain of f where $f'(c) = 0$ (indicating horizontal tangent lines) or $f'(c)$ does not exist (indicating vertical tangent lines or sharp points).

The First Derivative Test for Local Extrema

To determine the local extrema for a nonconstant function f continuous on the interval (a, b):

1. Find the _____ , x-values c in (a, b) such that $f'(c) = 0$ or $f'(c)$ is undefined.

2. Draw an _____ , and mark all critical x-values on it. Check the _____ of $f'(x)$ on each side of $x = c$. If the sign changes,

    i. from $+$ to $-$, then $f(c)$ is a local _____ .

    ii. from $-$ to $+$, then $f(c)$ is a local _____ .

3. If there is no sign change for $f'(x)$ from the left side of $x = c$ to the right side, then $f(c)$ is not a local extremum.

Example 2: Using the First Derivative Test

Let $f(x) = (x - 2)^3 + 1$.

a. Find the critical values of f.

b. Use the First Derivative Test to find any local extrema.

c. Sketch the graph of f.

Solution

Example 3: Critical Values

Find the critical values for the function $y = \dfrac{1}{x-2}$.

Solution

Example 4: Critical Values

Find the critical values for the function $y = x^{\frac{2}{3}}$.

Solution

11.5 Exercises

1. Consider the function:

$$f(x) = x^3 + 3x^2 - 105x + 1$$

    a. Find the critical values of the function. Separate multiple answers with commas.

    b. Use the First Derivative Test to find any local extrema. Write any local extrema as an ordered pair.

2. Consider the function:

$$f(x) = 9x + \frac{100}{x}$$

    a. Find the critical values of the function. Separate multiple answers with commas.

    b. Use the First Derivative Test to find any local extrema. Write any local extrema as an ordered pair.

3. Consider the function:

$$f(x) = \frac{4x^2 + 81}{x}$$

    a. Find the critical values of the function. Separate multiple answers with commas.

    b. Use the First Derivative Test to find any local extrema. Write any local extrema as an ordered pair.

4. A producer of clay repair kits has determined that the revenue from the production and sale of x units is given by $R(x) = 40x - 0.025x^2$ dollars, where $0 \le x \le 1300$. For what interval of production is the revenue increasing? For what interval is it decreasing? Express your answer in open intervals.

5. The revenue from the sale of x mineral-based makeup sets is given by $R(x) = 56x - 0.4x^2$ dollars. The total cost is given by $C(x) = 0.1x^2 + 730$ dollars, where $0 \le x \le 90$. Determine the interval of sales for which the profit is increasing and the interval for which it is decreasing. Express your answer in open intervals.

6. A typical student in an intermediate court reporting class can reach a level of recording

$$W(t) = 42 + \frac{53t^2}{t^2 + 20}$$ words per minute after t hours of instruction and practice.

a. Determine the interval for which the number of words recorded is increasing and the interval for which it is decreasing. Please express your answers as open intervals.

b. Over time, how many words per minute is the student's level of recording approaching?

11.6 Absolute Maximum and Minimum

Absolute Extrema

If c is in the domain of f and, for all x in the domain of f, we have

1. $f(x) \leq f(c)$, then $f(c)$ is called the _____ of f.

2. $f(x) \geq f(c)$, then $f(c)$ is called the _____ of f.

Theorem of Absolute Extrema of a Continuous Function

If a function f is continuous on a _____ interval $[a, b]$, then f will have an absolute maximum value and an absolute minimum value on $[a, b]$.

Finding the Absolute Extrema of a Continuous Function f on the Closed Interval $[a, b]$

1. Find all the _____ for f in $[a, b]$. That is, find all c in $[a, b]$ where

   a. $f'(c) = 0$ or

   b. $f'(c)$ is undefined.

2. Evaluate $f(a)$, $f(b)$, and $f(c)$ for all critical values c.

3. The _____ value found in step 2 is the absolute maximum. The _____ value found in step 2 is the absolute minimum.

Example 1: Finding Absolute Extrema

Find the absolute extrema for $f(x) = 2x^3 - 15x^2 + 24x + 6$ on the interval $[0, 5]$.

Solution

Example 3: Finding Absolute Extrema

Find the absolute extrema for $f(x) = x^{\frac{2}{3}} + 1$ on the interval $[-8, 8]$.

Solution

Example 5: Maximizing Profits

A company finds that its profit in dollars for producing x units of a product in one week is given by $P(x) = -2x^2 + 1600x$. If the company is set up so that no more than 500 units can be manufactured in any one week, how many units should the company produce to maximize profit?

Solution

11.6 Exercises

1. Consider the function $f(x) = 5x^2 - 20x$ on the interval $[0, 8]$. Find the absolute extrema for the function on the given interval. Express your answer as an ordered pair $(x, f(x))$.

2. Consider the function $f(x) = x^3 - 9$ on the interval $[-6, 7]$. Find the absolute extrema for the function on the given interval. Express your answer as an ordered pair $(x, f(x))$.

3. Consider the function $f(x) = 2x^3 + 6x^2 - 18x$ on the interval $[-7, 6]$. Find the absolute extrema for the function on the given interval. Express your answer as an ordered pair $(x, f(x))$.

4. Consider the function $f(x) = -4x^2 + 32x + 3$ on the interval $[-3, 9]$. Find the absolute extrema for the function on the given interval. Express your answer as an ordered pair $(x, f(x))$.

5. Consider the function $f(x) = 3x - 4$ on the interval $[-8, 5]$. Find the absolute extrema for the function on the given interval. Express your answer as an ordered pair $(x, f(x))$.

6. Consider the function $f(x) = 2x^{\frac{1}{2}} - \frac{\sqrt{2}}{6}x$ on the interval $[11, 29]$. Find the absolute extrema for the function on the given interval. Express your answer as an ordered pair $(x, f(x))$. (Round your answers to 3 decimal places.)

7. Consider the function $f(x) = -6x - \frac{486}{x}$ on the interval $[2, 11]$. Find the absolute extrema for the function on the given interval. Express your answer as an ordered pair $(x, f(x))$. (Round your answers to 3 decimal places.)

8. The weekly revenue from the sale of x units of a service is given by $R(x) = 104x - 4x^2$ thousand dollars, where $0 \le x \le 25$. How many units should be sold to maximize the revenue?

9. The weekly revenue from the production and sale of x units of salt is given by $R(x) = 205x - 2x^2$ thousand dollars. The cost function is given by $C(x) = 2x^2 + 53x + 2$ thousand dollars. Find the number of units of salt that are to be produced to maximize the profit if $0 \le x \le 25$.

10. On a good day, the contamination level in the air is approximately $f(x) = 0.13 + \dfrac{150x}{2.4x^2 + 60}$ ppm (parts per million), where x is the number of hours after 8:00 AM and $0 \le x \le 16$. How many hours after 8:00 AM will the contamination level reach its maximum?

CHAPTER 12

Applications of the Derivative

12.1 Concavity and Points of Inflection

Notation for the Second Derivative

The _____ of the function $y = f(x)$ can be denoted by any of the following symbols:

$$y'', f''(x), f^{(2)}, \frac{d^2y}{dx^2}, f_{xx}, \text{ and } D_x^2[y].$$

Example 1: First and Second Derivatives

Find both the first and second derivatives of the function $f(x) = x^3 - 12x + 1$.

Solution

Example 2: First and Second Derivatives

Find $\dfrac{dy}{dx}$ and $\dfrac{d^2y}{dx^2}$ if $y = \sqrt{x^2 + 1}$.

Solution

Example 3: First and Second Derivatives

For the function $f(u) = \dfrac{u^2}{u^2 - 9}$, find $f'(u)$ and $f''(u)$.

Solution

Concavity

Suppose that f is differentiable on the interval (a, b).

1. If f' is increasing on (a, b), then the graph of f is _____ on (a, b).

2. If f' is decreasing on (a, b), then the graph of f is _____ on (a, b).

Interpreting $f''(x)$ to Determine Concavity

Suppose that f is a function and f' and f'' both exist on the interval (a, b).

1. If $f''(x)$ _____ 0 for all x in (a, b), then f' is increasing and f is **concave** _____ on (a, b).

2. If $f''(x)$ _____ 0 for all x in (a, b), then f' is decreasing and f is **concave** _____ on (a, b).

Example 4: Determining Concavity

In the figure shown, the graph of $y = f(x)$ is given.

a. List the intervals on which f is concave upward.

b. List the intervals on which f is concave downward.

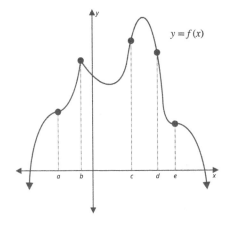

Solution

Point of Inflection

If the graph of a continuous function has a tangent line at a point (possibly a vertical line) and the graph changes concavity at that point, then the point is called a _____ .

To Determine Points of Inflection

Suppose that f is a continuous function.

1. Find $f''(x)$.

2. Find the hypercritical values of f. That is, find the values $x = c$ where

   a. $f''(c)$ _____ , or

   b. $f''(c)$ is _____ .

3. Using the hypercritical values as endpoints of intervals, determine the intervals where

   a. $f''(x)$ _____ 0 and f is concave upward, and

   b. $f''(x)$ _____ 0 and f is concave downward.

4. Points of inflection occur at those hypercritical values where f _____ concavity.

Example 6: Finding Points of Inflection

For the function $f(x) = x^3 - \dfrac{3}{2}x^2 + 5$,

   a. Determine the intervals on which f is concave upward and the intervals on which it is concave downward.

   b. Locate any points of inflection.

Solution

12.1 Exercises

1. Consider the function:

$$f(x) = -2x^2 - 5\sqrt[3]{x} - 9$$

Find $f''(x)$.

2. Consider the function:

$$f(x) = \frac{2x + 3}{-7x + 5}$$

Find $f''(x)$.

3. Consider the function:

$$f(x) = 5x^3 - 15x^2 - 8x - 2$$

    a. Determine the intervals on which the function is concave upwards or concave downwards.

    b. Locate any points of inflection. Write your answer as (x, y)-pairs.

4. Consider the function:

$$f(x) = -3x^3 - 288\sqrt{x} + 9$$

    a. Determine the intervals on which the function is concave upwards or concave downwards.

    b. Locate any points of inflection. Write your answer as (x, y)-pairs.

5. Consider the function:

$$f(x) = \sqrt[4]{10x^2 + 20}$$

   a. Determine the intervals on which the function is concave upwards or concave downwards.

   b. Locate any points of inflection. Write your answer as (x, y)-pairs.

6. Consider the function:

$$f(x) = \frac{-5x}{6x^2 - 294}$$

   a. Determine the intervals on which the function is concave upwards or concave downwards.

   b. Locate any points of inflection. Write your answer as (x, y)-pairs.

12.2 The Second Derivative Test

Second Derivative Test for Local Extrema

Suppose that f is a function, f' and f'' exist on the interval (a, b), c is in (a, b), and $f'(c) = 0$.

1. If $f''(c) > 0$, then $f(c)$ is a local _____ . (See Figure 4.1.8(a).)

2. If $f''(c) < 0$, then $f(c)$ is a local _____ . (See Figure 4.1.8(b).)

3. If $f''(c) = 0$, then the test _____ to give any information about local extrema.

Example 2: Locating Local Extrema

Locate the local extrema for the function $f(x) = \dfrac{8}{3}x^3 - x^4$ and sketch a graph of the function.

Solution

Point of Diminishing Returns

The **point of diminishing returns** occurs at a point of inflection where the sales curve changes from concave _____ to concave _____ .

Example 3: Point of Diminishing Returns

Find the point of diminishing returns for the sales function

$$S(x) = -0.02x^3 + 3x^2 + 100,$$

where x represents thousands of dollars spent on advertising, $0 \leq x \leq 80$, and S is sales in thousands of dollars for automobile tires.

Solution

12.2 Exercises

1. Use the Second Derivative Test to find all local extrema, if the test applies. Otherwise, use the First Derivative Test.

$$f(x) = -7x^2 + 126x - 560$$

2. Use the Second Derivative Test to find all local extrema, if the test applies. Otherwise, use the First Derivative Test.

$$f(x) = -6x^3 + 27x^2 - 36x$$

3. Consider the function:

$$f(x) = x^2 - 4\sqrt{x} + 2$$

   a. Find the second derivative of the given function.

   b. Use the Second Derivative Test to locate any local maximum or minimum points in the graph of the given function.

4. Consider the function:

$$f(x) = \sqrt{x^2 + 49}$$

    a. Find the second derivative of the given function.

    b. Use the Second Derivative Test to locate any local maximum or minimum points in the graph of the given function.

5. Use the Second Derivative Test to find all local extrema, if the test applies. Otherwise, use the First Derivative Test.

$$f(x) = 2x^3 - 9x^2 + 12x - 30$$

6. The cost function for a particular product is given by $C(x) = 0.001x^3 - 0.12x^2 + 23.3x + 180$ dollars, where $0 \leq x \leq 80$. Find the minimum marginal cost of the product, rounded to the nearest cent.

7. Consider the function:

$$f(x) = \left(x^2 - 9\right)^2$$

    a. Find the second derivative of the given function.

    b. Use the Second Derivative Test to locate any local maximum or minimum points in the graph of the given function.

12.3 Curve Sketching: Polynomial Functions

Example 1: Curve Sketching

Sketch a continuous function that satisfies all the given conditions.

List of Given Conditions
1. $f'(x) > 0$ for all x.
2. $f''(3) = 0$ and $f(3) = 4$.
3. $f''(x) < 0$ on $(-\infty, 3)$.
4. $f''(x) > 0$ on $(3, +\infty)$.

Solution

Example 2: Curve Sketching

Sketch a continuous function that satisfies all the given conditions.

List of Given Conditions
1. $g(-1) = 2$, $g(0) = 1$, $g(1) = -2$.
2. $g'(0) = 0$, $g'(1) = 0$.
3. $g''(0) = 0$, $g(.5) = 0$
4. $g''(x) < 0$ if $0 < x < .5$
5. $g''(x) > 0$ if $x < 0$ or $x > .5$.

Solution

Strategy for Curve Sketching Given a Function f

To sketch the graph of a function f:

1. Find $f'(x)$ and $f''(x)$.

2. Find the _____ of f. That is, find the values $x = c$ where

    a. $f'(c) = 0$, or

    b. $f'(c)$ is undefined.

3. Using the critical values as endpoints, find intervals

    a. where $f'(x)$ _____ 0 (f is increasing), and

    b. where $f'(x)$ _____ 0 (f is decreasing).

4. Locate the _____ using the First Derivative Test or the Second Derivative Test.

5. Find the _____ of f. That is, find the values $x = c$ where

    a. $f''(c) = 0$, or

    b. $f''(c)$ is undefined.

6. Using the hypercritical values as endpoints, find intervals

    a. where $f''(x)$ _____ 0 (f is concave upward), and

    b. where $f''(x)$ _____ 0 (f is concave downward).

7. Locate all _____. These occur at hypercritical values where the curve changes concavity.

8. Using the combined information from Steps 1 – 7 and any other specific points that might be helpful, sketch the graph. (A helpful point would usually be the y-intercept $(0, f(0))$ or an x-intercept: $(a, f(a))$ if $f(a) = 0$).

Example 4: Using the Strategy for Curve Sketching

Use the curve sketching strategy to sketch the graph of the function

$$f(x) = \frac{1}{3}x^3 - 2x^2 + 3x + 1.$$

Solution

Example 5: Graphing the Derivative

For the function graphed below:

a. Identify the local extrema and locate the point(s) of inflection.

b. Determine the intervals on which $f(x)$ is increasing and on which $f(x)$ is decreasing, and identify the intervals on which $f(x)$ is concave upwards and concave downwards.

c. Sketch on the same coordinate axis a possible graph of $f'(x)$.

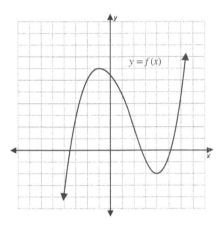

Solution

12.3 Exercises

1. Use the given conditions to determine which graph is a possible graph of the function $f(x)$.

    a. $f(1) = 2$

    b. $f'(1) = 0$

    c. $f'(x) < 0$ if $x < 1$

    d. $f'(x) > 0$ if $x > 1$

    e. $f''(x) > 0$ for all x

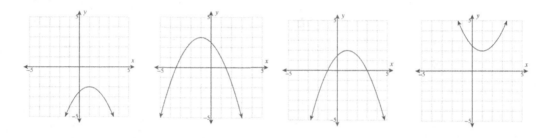

2. Consider the following function:

$$f(x) = x^4 - 18x^2 + 13$$

    a. Determine $f'(x)$ and $f''(x)$.

    b. Determine where the function is increasing and decreasing. Write your answers in interval notation.

    c. Determine where the function is concave up and concave down. Write your answers in interval notation.

    d. Identify the x-values of any local minima, maxima, and inflection points.

3. Consider the following function:

$$f(x) = -\frac{1}{2}x^4 + 2x^3 + 11$$

   a. Determine $f'(x)$ and $f''(x)$.

   b. Determine where the function is increasing and decreasing. Write your answers in interval notation.

   c. Determine where the function is concave up and concave down. Write your answers in interval notation.

   d. Identify the x-values of any local minima, maxima, and inflection points.

4. Suppose that $g(x) = \frac{1}{3}x^3 + sx^2 - 4x - 14$ has an inflection point at $x = 5$. Determine the value of s.

12.4 Curve Sketching: Rational Functions

Rational Function

A _____ is a function of the form

$$f(x) = \frac{P(x)}{Q(x)},$$

where $P(x)$ and $Q(x)$ are polynomials and $Q(x) \neq 0$.

Asymptotes for Rational Functions $f(x) = \dfrac{P(x)}{Q(x)}$

1. _____ are of the form $x = a$ and occur where the denominator is 0 and the numerator is not 0, (that is, $Q(a) = 0$ and $P(a) \neq 0$).

2. _____ are of the form $y = b$ and occur where

$$\lim_{x \to +\infty} f(x) = b \quad \text{or} \quad \lim_{x \to -\infty} f(x) = b.$$

3. _____ are of the linear form $y = mx + b$ and occur when the numerator is one degree larger than the denominator and $f(x)$ can be written as

$$f(x) = mx + b + \frac{R(x)}{Q(x)},$$

where

$$\lim_{x \to +\infty} \frac{R(x)}{Q(x)} = 0 \quad \text{or} \quad \lim_{x \to -\infty} \frac{R(x)}{Q(x)} = 0.$$

Example 1: Locating Vertical and Horizontal Asymptotes

For the function $f(x) = \dfrac{2x}{x - 5}$, find

a. the vertical asymptotes and

b. the horizontal asymptotes.

Solution

Example 2: Locating Vertical and Horizontal Asymptotes

For the function $f(x) = \dfrac{1}{x^2 + 3}$, find

a. the vertical asymptotes and

b. the horizontal asymptotes.

Solution

Example 3: Locating Oblique Asymptotes

Find the oblique asymptote for the function $f(x) = \dfrac{2x^2 + 7}{3x}$.

Solution

Example 5: Sketching the Graph of a Rational Function

Sketch the graph of the rational function $f(x) = \dfrac{100 + x^2}{2x}$.

Solution

12.4 Exercises

1. Consider the following rational function.

$$f(x) = \frac{-3x + 9}{-3x^2 - 8x + 16}$$

    a. Find the vertical asymptote(s), if any, of this function.

    b. Find the horizontal asymptote(s), if any, of this function.

    c. Find the oblique asymptote(s), if any, of this function.

2. Consider the following rational function.

$$f(x) = \frac{-2x^2 + 3x + 9}{2x - 3}$$

    a. Find the vertical asymptote(s), if any, of this function.

    b. Find the horizontal asymptote(s), if any, of this function.

    c. Find the oblique asymptote(s), if any, of this function.

3. Consider the following rational function:

$$f(x) = \frac{5}{4x - 8}$$

    a. Determine on which intervals the graph of the given function will be increasing and decreasing.

    b. Determine on which intervals the graph of the given function is concave up and concave down.

    c. Determine any local maxima, minima, and inflection points for the graph of the given function.

    d. Determine any vertical, horizontal, and oblique asymptotes for the given function.

12.5 Business Applications

Example 2: Maximizing Revenue

Susan knows that she can sell 440 tacos a day at 50 cents per taco. What price should she charge to maximize her revenue if, for every increase of 10 cents in price, she sells 40 fewer tacos? What will this revenue be?

Solution

Example 4: Finding Maximum Profit With a Linear Demand Function

Suppose that the company in Example 3 has a cost function $C(x) = 500 + 3x$. How many calculators should the company produce and sell to maximize profit?

Solution

12.5 Exercises

1. A gift shop expects to sell 400 wind chimes during the next year. It costs $1.50 to store one wind chime for one year. There is a fixed cost of $12 for each order. Find the lot size and the number of orders per year that will minimize inventory costs.

2. A beauty supply store expects to sell 110 flat irons during the next year. It costs $1.20 to store one flat iron for one year. There is a fixed cost of $16.50 for each order. Find the lot size and the number of orders per year that will minimize inventory costs.

3. A sporting goods store expects to sell 120 tennis rackets during the next year. It costs $4 to store one tennis racket for one year. To reorder, there is a fixed cost of $15, plus $0.90 for each tennis racket ordered. In what lot size and how many times per year should an order be placed to minimize inventory costs?

4. A furniture store expects to sell 120 coffee tables during the next year. It costs $1.50 to store one coffee table for one year. To reorder, there is a fixed cost of $10, plus $2.80 for each coffee table ordered. In what lot size and how many times per year should an order be placed to minimize inventory costs?

5. A jewelry store sells 216 diamond bracelets per month at $1100 each. The owners estimate that for each $55 increase in price, they will sell 6 fewer diamond bracelets per month. Find the price per diamond bracelet that will maximize revenue.

6. A bookstore sells 160 calculus textbooks per month at $130 each. The owners estimate that for each $13 increase in price, they will sell 5 fewer calculus textbooks per month. Find the price per calculus textbook that will maximize revenue.

12.6 Other Applications: Optimization, Distance, and Velocity

Example 1: Volume of a Box

A rectangular box with no top is to be made from a rectangular piece of cardboard that is 18 in. long by 12 in. wide. Small squares (all the same size) are to be cut from each corner and the sides folded to form the box. What size should the squares be to give a maximum volume to the box?

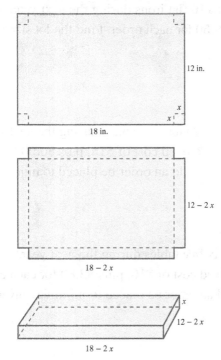

Solution

Example 2: Surface Area of a Box

A rectangular box is to be made so that the top and bottom are squares. The volume is to be 250 cm³. Material for the top and bottom costs $2 per square centimeter, but the material for the sides costs only $1 per square centimeter. What dimensions will give a minimum cost for the materials? What is the minimum cost?

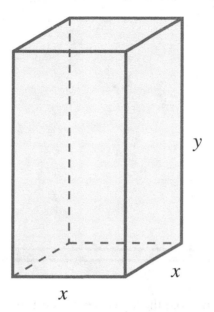

Solution

Example 3: Fencing

A farmer wants to enclose a rectangular area next to his barn. If he has 200 feet of fencing to use, what dimensions of the rectangle will maximize the area? What is the maximum area? (Note that the fence is only three sides of the rectangle, since the barn serves as the fourth side of the enclosure.)

Solution

Example 6: Vertical Projectile

A projectile is fired vertically (straight up) from the top corner of the 176-foot tall building. The height of the projectile at time t (in seconds) is given by the function $s(t) = -16t^2 + 160t + 176$.

a. What is the maximum height of the projectile?

b. When does the projectile hit the ground?

c. How fast is the projectile moving when it hits the ground?

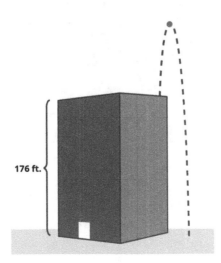

176 ft.

Solution

12.6 Exercises

1. A shipping company must design a closed rectangular shipping crate with a square base. The volume is 36864 ft^3. The material for the top and sides costs $3 per square foot and the material for the bottom costs $13 per square foot. Find the dimensions of the crate that will minimize the total cost of material.

2. A brick wall makes two legs of a right angle, one 21 feet long and the other 33 feet long. A contractor is told to add 102 feet of brick fencing to complete a rectangular fence. Determine the maximum possible enclosed area by finding the dimensions of the completed fence that will obtain this area.

3. An environmental landscaping company wishes to run a pipeline from a pumping platform (R) located 15 miles offshore to a terminal (B) 19 miles down the coast. It costs $153,000 per mile to lay the pipeline underwater and $72,000 per mile to lay the pipeline over land. Determine how many miles from the terminal the two types of pipe should meet (P) so that the total cost is minimized. Write the exact answer or round to the nearest hundredth.

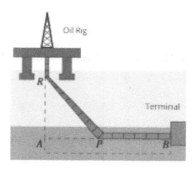

4. A man (M) is standing on the bank of a river that is 0.6 miles wide. He wants to reach a house (B) on the opposite shore that is 1.7 miles downstream. The man can row the boat 2 mph and can walk 4 mph. Find the distance between the house and the point (P) where he should dock his boat in order to minimize the total time he would need to reach the house. Write the exact answer or round to the nearest hundredth.

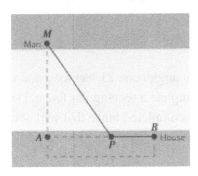

5. A projectile is fired vertically, and its height (in feet) after t seconds is given by $s(t) = -5t^2 + 70t + 7755$.

    a. Find the maximum height of the projectile. Round your final answer to the nearest hundredth.

    b. At what time will the projectile hit the ground? Round your final answer to the nearest hundredth.

    c. At what rate is the projectile falling when it hits the ground? Round your final answer to the nearest hundredth.

6. A rectangular box with no top is to be made from a piece of sheet metal that is 63 ft by 135 ft in size. Equal squares are to be cut from each corner and the sides folded to form the box.

What is the side length of the squares that are to be cut out of the corners if the box is to have maximum volume? Write the exact answer or round to the nearest hundredth.

7. A rectangular box is designed to have a square base and an open top. The volume is to be 5324 in.^{3}

What is the minimum surface area that such a box can have?

8. A farmer wants to build a rectangular pen and then divide it with two interior fences. The total area inside of the pen will be 2196 square meters. The exterior fencing costs $17.28 per meter and the interior fencing costs $12.00 per meter.

Find the dimensions of the pen that will minimize the cost.

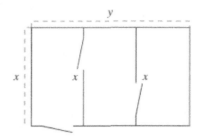

CHAPTER 13

Additional Applications of the Derivative

13.1 Derivatives of Logarithmic Functions

Derivative of $\ln x$

If $f(x) = \ln x$, then

$$\underline{\hspace{5cm}}.$$

Example 2: Finding the Derivative

Find $\dfrac{dy}{dx}$ for $y = x^2 \ln x$.

Solution

Derivative of $\ln(g(x))$

If $y = \ln(g(x))$, then

$$\underline{\hspace{7cm}}.$$

Example 4: Finding the Derivative

a. Find $f'(x)$ for $f(x) = \ln(x^2 + 2x + 10)$.

Solution

b. Find $f'(x)$ for $f(x) = \ln \sqrt{x^2 + 3}$.

Solution

Example 5: Using Logarithmic Differentiation

Use logarithmic differentiation to find $f'(x)$, given that $f(x) = \dfrac{(x^2+1)^3 \sqrt{x^2 - 2x}}{(x-5)}$.

Solution

Example 7: Graphing Logarithmic Functions

Using curve sketching techniques, sketch the graph of the function $f(x) = x \ln x$.

Solution

13.1 Exercises

1. Find the absolute extrema for the given function on the interval $[0.58, 4.5]$. Write your answer in the form $(x, f(x))$. Round your answers to two decimal places.

$$f(x) = 3x^2 - 6\ln(x^4)$$

2. A carpenter has determined that the profit from the sale of x benches is given by the function $P(x) = 5x + \ln(8x^3 + 11)$ dollars. Find the marginal profit function $P'(x)$.

3. A furniture dealer has estimated that he can sell $N(x) = 170 + 57\ln(2 + 1.5x)$ pieces of furniture annually, where x (in thousands of dollars) is the amount spent on advertising.

   a. Find the number of pieces of furniture sold if \$8000 is spent on advertising. Round your answer to the nearest integer.

   b. Find the rate of change in the number of pieces of furniture sold if \$8000 is spent on advertising. Round your answer to the nearest integer.

4. Find the derivative of the given expression.

$$f(x) = 5x^3 \ln(x^5)$$

5. Find the derivative of the given expression.

$$f(x) = \frac{\ln(x^4)}{x^3}$$

6. Find the derivative of the given expression.

$$f(x) = \ln\left(\left(10x^2 + 17x\right)^2\right)$$

7. Find the derivative of the given expression.

$$f(x) = \frac{\ln\left(17x^2 - 23x^3\right)}{4x^5}$$

8. Find the derivative of the given expression.

$$f(x) = \ln\left(\left(6x^3 - 22x\right)^4 \left(14x - 13x^2\right)\right)$$

9. Find the derivative of the given expression.

$$y = \ln\left(\left(\frac{9 + 8x^2}{14x^2 + 8}\right)^2\right)$$

13.2 Derivatives of Exponential Functions

Derivative of the Exponential Function

If $f(x) = e^x$, then

$$\underline{\hspace{4cm}}.$$

Example 1: Finding the Derivative

a. Find $\dfrac{dy}{dx}$ for $y = x^3 + e^x$.

Solution

b. Find $\dfrac{dy}{dx}$ for $y = x^2 e^x$.

Solution

Chain Rule for Exponential Functions

If $f(x) = e^{g(x)}$, then

$$\underline{\hspace{5cm}}.$$

Example 2: Using the Chain Rule

Find $\dfrac{dy}{dx}$ if $y = e^{x^2 + 3x}$.

Solution

Example 4: Graphing Exponential Functions

For $f(x) = xe^{-x}$

a. Find any critical values.

b. Find any hypercritical values.

c. Sketch the graph of the function.

Solution

13.2 Exercises

1. Consider the following function:

$$g(x) = -4x^2 e^x$$

    a. Find the first derivative of the above function.

    b. Find the second derivative of the above function.

2. Find a formula for $f'(x)$ and determine the slope $f'(2)$ for the following function.

$$f(x) = \frac{-e^x}{5e^x + 8}$$

3. Find a formula for y' and determine the slope $y'|_{x=2}$ for the following function.

$$y = e^{5-x^5} \ln(x^2 + 2)$$

4. Find a formula for y' and determine the slope $y'|_{x=-1}$ for the following function.

$$y = -8e^{3x^3+2}$$

5. Consider the following function:

$$g(x) = e^{-0.8x^2+1}$$

  a. Find the first derivative of the given function.

  b. Use $g'(x)$ to determine the intervals on which the given function is increasing or decreasing. Separate multiple intervals with commas.

6. Consider the following function:

$$f(x) = 3(8e^{4x} + 6)^3$$

  a. Find the first derivative of the given function.

  b. Use $f'(x)$ to determine the intervals on which the given function is increasing or decreasing. Separate multiple intervals with commas.

7. Find the absolute extrema of the function over the given interval. Write your answers as ordered pairs. Round your final answer to four decimals.

$$g(x) = 8x^3 e^{-0.8x^2}; [-5, 2]$$

13.3 Growth and Decay

Exponential Growth

A function $y = f(t)$ satisfies the equation

$$\frac{dy}{dt} = ky,$$

if and only if

$$y = ce^{kt}.$$

If k and c are positive, we say that y _____ (or **increases**) _____ .

Example 1: Population Growth

Suppose that a population of rabbits grows exponentially; that is, the rate of population growth depends on the size of the population. Further, suppose that on Monday at 8 A.M. there are 100 rabbits and that at 8 A.M. on Wednesday there are 130 rabbits.

   a. Find an exponential function that represents the rabbit population at time t, where t is measured in days from Monday at 8 A.M.

   b. Find the time that the population will be 200 rabbits (double the original 100).

Solution

Example 2: Continuous Compounding Interest

Suppose that $2000 is invested in an account that earns 8 percent compounded continuously.

a. What will be the balance in 1 year?

b. When will the original investment be doubled?

Solution

Exponential Decline

A function $y = f(t)$ satisfies the equation

$$\frac{dy}{dt} = -ky,$$

if and only if

$$y = ce^{-kt}.$$

If k and c are positive, we say that y _____ (or **decays**) _____ .

Limited Growth

The mathematical model for _____ is

$$y = c\left(1 - e^{-kt}\right),$$

where c and k are positive constants.

Example 5: Limited Growth

Suppose that the decay of a particular radioactive isotope is described by the equation

$$y = 200e^{-0.05t},$$

where y is the amount of the isotope in grams and t is time in years. Find the half-life of this isotope.

Solution

General Exponential Derivative

1. If $f(x) = B^x$, then $f'(x) = \ln B \cdot f(x) =$ _____.

2. Chain Rule: if $f(x) = B^{g(x)}$
     then $f'(x) =$ _____.

General Logarithmic Derivative

1. If $y = \log_B(x)$ then

_____.

2. Chain Rule: If $y = \log_B(g(x))$, then

_____.

Example 6: Using the Derivatives

Given $f(x) = (x^2 - 3x + 6) \log_2(x)$, determine $f'(x)$.

Solution

13.3 Exercises

1. A population of rabbits grows exponentially at a rate of 3.3 percent per year. The population was 2900 in 1996.

   a. Find the exponential function that represents the population t years after 1996.

   b. What will the population be in the year 2001? Round up to the nearest whole animal.

   c. In what year will the population be 6800? Round to the nearest year.

2. In 1997, the cost of a rare coin was about $8. In 2001, the cost was $20. If the cost is growing exponentially, predict the cost of the rare coin in 2012. Round to the nearest cent.

3. One thousand dollars is deposited in a savings account where the interest is compounded continuously. After 8 years, the balance will be $1452.12. When will the balance be $1611.15? Round to the nearest tenth of a year.

4. The amount of goods and services that costs $600 on January 1, 2000 costs $693.94 on January 1, 2008. Estimate the cost of the same goods and services on January 1, 2011. Assume the cost is growing exponentially. Round your answer to the nearest cent.

5. The decay rate for a radioactive isotope is 3.6 percent per year. Find the half-life of the isotope. Round to the nearest tenth of a year.

6. Studies have shown that the fractional part P of a particular brand of batteries that has died after t hours of continuous use is given by the function $P(t) = 1 - e^{-0.05t}$.

   a. What percentage of the batteries has died after 14 hours? Round your answer to the nearest percent.

   b. How long will it be before $\dfrac{1}{4}$ of the batteries are dead? Round your answer to the nearest tenth of an hour.

7. The time that it takes a carpenter to build a shelving unit is given by the function $T(x) = 37 + ce^{-kx}$ minutes, where x is the number of units that the carpenter has made before. It takes the carpenter 47 minutes to build the first shelving unit $(x = 0)$ and 43 minutes to build the tenth unit. How long will it take the carpenter to build the nineteenth unit? Round your answer to the nearest tenth of a minute.

13.4 Elasticity of Demand

Elasticity of Demand

If $p = D(x)$ is the demand function for a product, then the _____ for

that product is

$$E = -\frac{1}{x} \cdot \frac{D(x)}{D'(x)}.$$

Properties of Elasticity of Demand

1. If $E > 1$, then $R'(x) = D(x)\left[1 - \frac{1}{E}\right] > 0$, the revenue is _____ , and we say that

the demand is _____ .

2. If $E < 1$, then $R'(x) = D(x)\left[1 - \frac{1}{E}\right] < 0$, the revenue is _____ , and we say that

the demand is _____ .

3. If $E = 1$, then $R'(x) = 0$, the revenue is at its _____ , and we say that the demand has

_____ .

Example 2: Elasticity of Demand

A product is known to have a demand function $p = D(x) = 100e^{-0.5x}$.

a. Find the value of x for which $E = 1$.   Note: these should be the same x-value.

b. Find the value for x that maximizes the revenue.

Solution

Example 3: Alternative Approach to Elasticity of Demand

A grocery store determines that the demand function for its bakery's bread is $x = 180 - 30p$, where x is the number of loaves of bread it sells daily and p is the unit price.

a. Find the quantity demanded when the price is $1.70.

b. Find the function describing the elasticity as a function of p.

c. Find the elasticity at $p = \$1.70$.

d. Interpret the resulting elasticity.

e. Determine the revenue function R and find p so that R is a maximum.

Solution

13.4 Exercises

1. Consider the following demand function:

$$p = D(x) = 114 - 1.9x$$

a. Find the elasticity function.

b. Find the value of x that maximizes the revenue.

2. Consider the following demand function:

$$p = D(x) = \sqrt{148.5 - 1.5x}$$

    a. Find the elasticity function.

    b. Find the value of x that maximizes the revenue.

3. Consider the following demand function:

$$p = D(x) = 12 - 2\sqrt{x}$$

    a. Find the elasticity function.

    b. Find the value of x that maximizes the revenue.

4. The demand function for a stapler is given by $p = D(x) = 20.8 - 0.4x$ dollars. Find the level of production for which the revenue is maximized.

5. The demand function for a stapler is given by $p = D(x) = 3249 - 0.03x^2$ dollars. Find the level of production for which the revenue is maximized.

6. The demand function for a commodity is given by $p = D(x) = 74 - 6\sqrt{x}$ dollars. Find the level of production for which the demand is elastic. Remember that you can only produce whole units.

7. The demand function for a particular item is given by $p = D(x) = \sqrt{170 - 6x}$ dollars. Find the level of production for which the demand is elastic. Remember that you can only produce whole units.

13.5 L'Hôpital's Rule

Cauchy's Mean Value Theorem

Suppose that f and g are continuous on $[a, b]$ and differentiable on (a, b), $g'(x) \neq 0$ on (a, b), and $g(a) \neq g(b)$. Then there is a point $c \in (a, b)$ such that

$$\frac{f'(c)}{g'(c)} = \underline{\hspace{6cm}}$$

L'Hôpital's Rule

Suppose f and g are differentiable at all points of an open interval I containing c, and that $\overset{?}{g}(x) \neq 0$ for all $x \in I$ except possibly at $x = c$. Suppose further that either

$$\lim_{x \to c} f(x) = \underline{\hspace{2cm}} \text{ and } \lim_{x \to c} g(x) = \underline{\hspace{2cm}}$$

or

$$\lim_{x \to c} f(x) = \underline{\hspace{2cm}} \text{ and } \lim_{x \to c} g(x) = \underline{\hspace{2cm}}.$$

Then

$$\lim_{x \to c} \frac{f(x)}{g(x)} = \underline{\hspace{6cm}}$$

assuming the limit on the right is a real number or ∞ or $-\infty$.

Further, the rule is true for one-sided limits at c and for limits at infinity; that is, $x \to c$ can be replaced with $x \to c^+$, $x \to c^-$, $x \to -\infty$, or $x \to \infty$, assuming always that the limit on the right is a real number or ∞ or $-\infty$.

Caution!

L'Hôpital's Rule says the limit of a quotient of two functions is equal to the limit of the quotient of their derivatives—don't mistakenly apply the Quotient Rule of differentiation! After verifying that the conditions of l'Hôpital's Rule are satisfied, proceed by differentiating the $\underline{\hspace{3cm}}$ and $\underline{\hspace{3cm}}$ $\underline{\hspace{3cm}}$ individually.

Example 4

Evaluate the following limits.

a. $\displaystyle\lim_{x \to \infty} \frac{e^x}{5x^2 - 3x + 2}$

Solution

Example 5

Determine $\lim\limits_{x \to 0^+} \sqrt{x} \ln x$

Solution

Example 7

Determine $\lim\limits_{x \to 0^+} (1 + x)^{\frac{1}{x}}$

Solution

Example 8

Determine $\lim\limits_{x \to 0^+} x^x$

Solution

Example 9

Determine $\lim\limits_{x \to \infty} x^{\frac{1}{x}}$

Solution

13.5 Exercises

1. Check whether l'Hôpital's Rule applies to the given limit. If it does, use it to determine the value of the limit. If it does not, find the limit some other way. (If necessary, apply l'Hôpital's Rule several times.)

$$\lim\limits_{t \to 0} \frac{7t}{\sqrt{2t+9} - 3}$$

2. Check whether l'Hôpital's Rule applies to the given limit. If it does, use it to determine the value of the limit. If it does not, find the limit some other way. (If necessary, apply l'Hôpital's Rule several times.)

$$\lim\limits_{x \to \infty} \frac{9 \ln x}{\ln\left(5x^2 + 7x\right)}$$

3. If the following limit is of indeterminate form, use l'Hôpital's Rule to find it. If the limit is not of indeterminate form, find it by other means.

$$\lim\limits_{x \to 0^+} 10x \ln\left(2x\right)$$

4. Identify the indeterminate form for the following limit. If the limit is not of indeterminate form, write *Not indeterminate*.

$$\lim\limits_{x \to 0^+} 4x \ln\left(5x\right)$$

5. Identify the indeterminate form for the following limit. If the limit is not of indeterminate form, write *Not indeterminate*.

$$\lim_{x \to 0^+} \left(\frac{8}{x} \right)^{5x}$$

6. Identify the indeterminate form for the following limit. If the limit is not of indeterminate form, write *Not indeterminate*.

$$\lim_{x \to 1^+} \left(\frac{5}{\ln x} - \frac{9}{x-1} \right)$$

7. If the following limit is of indeterminate form, use l'Hôpital's Rule to find it. If the limit is not of indeterminate form, find it by other means.

$$\lim_{x \to 0^+} (1 - 3x)^{\frac{5}{x}}$$

8. Identify the indeterminate form for the following limit. If the limit is not of indeterminate form, write *Not indeterminate*.

$$\lim_{x \to 0^+} (1 - 8x)^{\frac{6}{x}}$$

13.6 Differentials

Differential

If $y = f(x)$ is a function and $y' = f'(x)$ exists, then the **differential** dy is defined as

$$dy = \underline{\hspace{4cm}}.$$

Example 1: Finding Differentials

b. Find du for $u = \left(x^2 + 5\right)^3$.

Solution

Example 3: Estimating Square Roots

Using differentials, estimate $\sqrt{15}$.

Solution

Example 4: Volume

The volume of a cube with edge x is given by $V = x^3$. Using differentials, approximate the change in V if x is

a. Changed from 5 cm to 5.01 cm.

b. Changed from 5 cm to 4.99 cm.

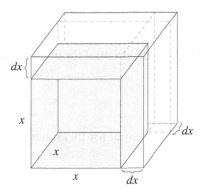

Solution

Example 5: Change in Revenue

Suppose that a company makes and sells x tennis rackets per week, and the corresponding revenue function (in hundreds of dollars) is $R(x) = 20x - \dfrac{x^2}{30}$, where x is between 0 and 600. Use dR to estimate the approximate change in revenue if production is

  a. Increased from 150 to 160 rackets.

  b. Increased from 480 to 490 rackets.

Solution

13.6 Exercises

1. Consider the function $u = (3t + 8)^2$. Find the differential for this function.

2. Consider the function $V = 9\pi r^3$. Find the differential for this function.

3. Consider the function $y = -7x^2 + \dfrac{8}{x^2}$. Using the values $x = -4$ and $\Delta x = -0.9$, calculate $\Delta y - dy$. Round your answer to three decimal places if necessary.

4. Consider the function $y = \sqrt{36x^2 + 36}$. Using the values $x = -4$ and $\Delta x = 0.7$, calculate $\Delta y - dy$. Round your answer to three decimal places if necessary.

5. Use differentials to approximate $\sqrt{19}$.

Express your answer as $a \pm \dfrac{b}{c}$ where a is the nearest integer.

6. Use differentials to approximate $\sqrt[3]{43}$.

Express your answer as $a \pm \dfrac{b}{c}$ where a is the nearest integer.

7. Use differentials to approximate $\sqrt{26.3}$.

Express your answer as $a \pm \dfrac{b}{c}$ where a is the nearest integer.

8. Use differentials to approximate $\sqrt[3]{3.4}$.

Express your answer as $a \pm \dfrac{b}{c}$ where a is the nearest integer.

9. A total cost function (in dollars) is given by $C(x) = 460 + 6x + 0.04x^2$. Use differentials to estimate the change in cost when the level of production is changed from 122 to 126 units.

10. Suppose that a company makes and sells x radios per week, and the corresponding revenue function is $R(x) = 745 + 49x + 0.11x^2$. Use differentials to estimate the change in revenue if production is changed from 169 to 165 units.

CHAPTER 14

Integration with Applications

14.1 The Indefinite Integral

Antiderivative

If F and f are two functions and _____ for all

_____ , then F is an **antiderivative** of f.

Note: We also say that $F(x)$ is an antiderivative of $f(x)$.

Antiderivatives Differ by a Constant

If G and F are both antiderivatives of _____ f, then

$$G(x) = \underline{\hspace{2cm}} \, ,$$

where C is a _____ .

Example 1: Antiderivatives

Show that both $F(x) = x^3 + x^2 + 50$ and $G(x) = x^3 + x^2 - 80$ are antiderivatives of $f(x) = 3x^2 + 2x$.

Solution

The Indefinite Integral

If $F(x)$ is any antiderivative of $f(x)$, then the **indefinite integral** of $f(x)$, symbolized by $\displaystyle\int f(x)dx$, is

defined as

$$\int f(x)dx = \underline{\hspace{2cm}} \, ,$$

where C is an arbitrary constant. $f(x)$ is called the _____ , dx is called the

_____ , and C is called the _____ .

Formulas of Integration

I. $\int k\,dx = $ _____

II. $\int x^r\,dx = $ _____ , where $r \neq -1$.

III. $\int \frac{1}{x}\,dx = \int x^{-1}\,dx = $ _____   Note: the power rule, formula II, does not apply.

IV. $\int e^x\,dx = $ _____

Constant Multiple Rule

$$\int k\,f(x)\,dx = $$
_____

In words, the integral of a constant times a function is equal to

_____ .

Sum and Difference Rule

$$\int [f(x) \pm g(x)]\,dx = $$
_____

In words, the integral of a sum (or difference) of two functions is the _____ (or

_____) of their _____ .

Example 2: Finding the Indefinite Integral

Find $\int 3x^5\,dx$.

Solution

Example 4: Finding the Indefinite Integral

Find $\int \dfrac{5}{x}\,dx$.

Solution

Example 7: Finding the Indefinite Integral of a Rational Function

Find $\int \dfrac{t^3 + t + 1}{t^2}\,dt$.

Solution

Example 8: Finding the Constant of Integration

If $F'(x) = x - \dfrac{1}{x}$ and $F(1) = \dfrac{3}{2}$, find $F(x)$.

Solution

14.1 Exercises

1. Find the following indefinite integral.

$$\int \left(-7x^6 - 5 \right) \, dx$$

2. Find the following indefinite integral.

$$\int \left(e^x + 10 \right) \, dx$$

3. Find the following indefinite integral.

$$\int \left(-\frac{8}{9}x^{-\frac{1}{2}} - 5 \right) \, dx$$

4. Find the following indefinite integral.

$$\int \left(\frac{-9}{x} - 7x^{-\frac{1}{3}} + 3e^x \right) \, dx$$

5. Find the following indefinite integral.

$$\int \left(-7\sqrt[3]{x} - 9\sqrt{x} \right) \, dx$$

6. Find the following indefinite integral.

$$\int \left(6e^x + 6x^4 + \frac{6}{7} \right) \, dx$$

7. Find the following indefinite integral.

$$\int \left(\frac{1}{\sqrt[4]{t}} - \frac{3}{t^8} \right) dt$$

8. Perform the indicated multiplication and then integrate.

$$\int (8x + 3)^2 \, dx$$

9. Simplify the indicated quotient and then integrate.

$$\int \frac{3x^6 - 5x^5 - 2}{x^6} \, dx$$

10. The weekly marginal cost of producing x cameras is $19 + 0.11x$ dollars per camera. Find the cost function if the fixed costs are $3000.

14.2 Integration by Substitution

Formulas of Integration with Function Notation

I. $\int kg'(x)dx = $ _____

II. $\int [g(x)]^r g'(x)dx = $ _____ , where $r \neq -1$

III. $\int \dfrac{g'(x)dx}{g(x)} = \int \dfrac{1}{g(x)}g'(x)dx = \int [g(x)]^{-1}g'(x)dx = $ _____

IV. $\int e^{g(x)}g'(x)dx = $ _____

Formulas of Integration with u-Substitution

I. $\int k\,du = $ _____

II. $\int u^r du = $ _____ , where $r \neq -1$

III. $\int \dfrac{du}{u} = \int \dfrac{1}{u}du = \int u^{-1}du = $ _____

IV. $\int e^u du = $ _____

Example 1: Integration by Substitution with Rational Expressions

Find $\int \dfrac{2t}{t^2+1}dt$.

Solution

Example 2: Integration by Substitution with Radicals

Find $\int \sqrt{5x + 1}\, dx$.

Solution

Example 4: Integration by Substitution

Find $\int \dfrac{x^2}{(x^3 - 1)^5}\, dx$.

Solution

Example 5: Incorrect Integration by Substitution

Find $\int 3x^2 \sqrt{x^3 + 1}\, dx$.

Solution

14.2 Exercises

1. Use integration by substitution to solve the integral below. Use C for the constant of integration.

$$\int (x + 2)^7 \, dx$$

2. Use integration by substitution to solve the integral below. Use C for the constant of integration.

$$\int \frac{1}{x + 2} \, dx$$

3. Use integration by substitution to solve the integral below. Use C for the constant of integration.

$$\int e^{2x^9 + 9} \cdot 18x^8 \, dx$$

4. Use integration by substitution to solve the integral below. Use C for the constant of integration.

$$\int -2(3y - 5)^{-2} \, dy$$

5. Use integration by substitution to solve the integral below. Use C for the constant of integration.

$$\int \frac{6}{\sqrt[3]{9t + 7}} \, dt$$

6. Use integration by substitution to solve the integral below. Use C for the constant of integration.

$$\int -5y e^{3y^2} \, dy$$

7. Use integration by substitution to solve the integral below. Use C for the constant of integration.

$$\int \frac{-6e^t}{e^t + 7} \, dt$$

8. Use integration by substitution to solve the integral below. Use C for the constant of integration.

$$\int -4e^{-4y} \left(7 - 4e^{-4y}\right)^2 \, dy$$

9. Use integration by substitution to solve the integral below. Use C for the constant of integration.

$$\int -5y^2 \sqrt{4y^3 - 1} \, dy$$

10. Use integration by substitution to solve the integral below. Use C for the constant of integration.

$$\int \frac{3(\ln(x))^3}{x} \, dx$$

14.3 Area and Riemann Sums

Example 1: Using the Fundamental Theorem of Calculus

Use the Fundamental Theorem of Calculus to determine the area under the graph of $f(x) = 30 - 20x + x^5$ and over the x-axis interval $[1, 2]$.

Solution

General Form of a Riemann Sum

The **general form of a Riemann sum** for a function $y = f(x)$ continuous on $[a, b]$ is

$$\underline{\hspace{8cm}},$$

where $\Delta x = \underline{\hspace{3cm}}$, n is the number of subintervals of $[a, b]$ and each of the numbers $c_1, c_2, \ldots, c_n$ represents one x-value from each subinterval.

Example 3: Riemann Sum

Use a Riemann sum to estimate the area between $y = x^2 + 1$ and the x-axis on the interval $[a, b] = [-1, 2]$ with $n = 6$.

Solution

14.3 Exercises

1. Find the total area of the rectangles in the figure below, where the equation of the line is $f(x) = -2x + 11$.

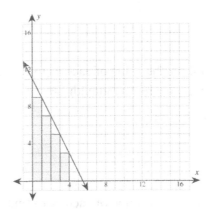

2. Find the total area of the rectangles in the figure below, where the equation of the curve is $f(x) = -x^2 + 15$.

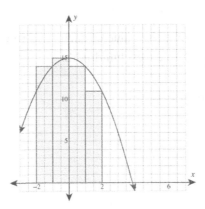

3. Find the Riemann sum S_4 for the following information. Round your answer to the nearest hundredth.

$$f(x) = 49 - x^2; \ [a, b] = [-7, 5]; \ n = 4, \ c_1 = -6.5, \ c_2 = -3.5, \ c_3 = -0.5, \ c_4 = 2.5$$

4. Find the Riemann sum S_4 for the following information. Round your answer to the nearest hundredth.

$$f(x) = \sqrt{9 - x}; \ [a, b] = [-9, -1]; \ n = 4, \ c_1 = -8.5, \ c_2 = -6.5, \ c_3 = -4.5, \ c_4 = -2.5$$

5. Find the Riemann sum S_4 for the following information. Round your answer to the nearest hundredth.

$$f(x) = \frac{1}{x+7}; [a,b] = [-6, -2]; n = 4, c_1 = -5.5, c_2 = -4.5, c_3 = -3.5, c_4 = -2.5$$

6. Use the midpoint of each subinterval for the value of each c_k to find the Riemann sum S_5 for the following information. Round your answer to the nearest hundredth.

$$f(x) = 64 - x^2; [a,b] = [-8, -3]; n = 5$$

7. Use the midpoint of each subinterval for the value of each c_k to find the Riemann sum S_4 for the following information. Round your answer to the nearest hundredth.

$$f(x) = \sqrt{7 - x}; [a,b] = [-7, -3]; n = 4$$

8. Use the midpoint of each subinterval for the value of each c_k to find the Riemann sum S_4 for the following information. Round your answer to the nearest hundredth.

$$f(x) = \frac{1}{x+6}; [a,b] = [-5, -1]; n = 4$$

14.4 The Definite Integral and the Fundamental Theorem of Calculus

Area Under a Curve Defined

If a function $y = f(x)$ is _____ on the interval $[a, b]$

then the **area under the curve** is defined to be

_____ ,

where S_n is the general form of a Riemann sum for the function f.

The Definite Integral

If a function $y = f(x)$ is continuous on the interval $[a, b]$, then the _____ **integral** of f from

a to b, symbolized as $\int_a^b f(x)dx$, is defined to be

_____ ,

where S_n is the general form of a Riemann sum for the function f.

The number a is called the _____ **limit** of integration and the number b is called the _____

limit of integration.

Formula for Area Under a Curve

For a function $y = f(x)$, nonnegative and continuous on $[a, b]$, the area under the curve is given by

_____ .

The Fundamental Theorem of Calculus

If a function $y = f(x)$ is continuous on the interval $[a, b]$ and $F(x)$ is any antiderivative of $f(x)$, then

$$\int_a^b f(x)dx = [F(x)]_a^b = \underline{\hspace{4cm}} .$$

Example 1: Definite Integrals

a. Evaluate $\displaystyle\int_1^2 2x\,dx$.

Solution

b. Find the value of $\displaystyle\int_1^4 \frac{1}{\sqrt{x}}\,dx$.

Solution

Properties of Definite Integrals

1. $\displaystyle\int_a^a f(x)\,dx = \underline{\qquad}$

2. $\displaystyle\int_a^b k\,f(x)\,dx =$

$\underline{\hspace{8cm}}$

3. $\displaystyle\int_a^b [f(x) \pm g(x)]\,dx =$

$\underline{\hspace{9cm}}$

Example 2: Using Properties of Definite Integrals

b. Find the value of $\displaystyle\int_0^1 x\sqrt{x^2+1}\,dx$.

Solution

Average Value

For a function $y = f(x)$, continuous on the interval $[a, b]$, the **average value** is

$$AV = \underline{\hspace{8cm}}\,.$$

Example 3: Average Value

a. Find the average value of $f(x) = 8x - x^2$ on the interval $[0, 8]$.

Solution

14.4 Exercises

1. Evaluate the definite integral below.

$$\int_{2}^{3} -x^3 \, dx$$

Write your answer in exact form or rounded to two decimal places.

2. Evaluate the definite integral below.

$$\int_{-5}^{-2} \frac{7}{\sqrt[3]{x^2}} \, dx$$

Write your answer in exact form or rounded to two decimal places.

3. Evaluate the definite integral below.

$$\int_{1}^{4} \left(2x^3 - 2\right) \, dx$$

Write your answer in exact form or rounded to two decimal places.

4. Evaluate the definite integral below.

$$\int_{6}^{13} \frac{6}{x} \, dx$$

Write your answer in exact form or rounded to two decimal places.

5. Evaluate the definite integral below.

$$\int_{1}^{4} \left(3 + 4e^x\right) \, dx$$

Write your answer in exact form or rounded to two decimal places.

6. Evaluate the definite integral below.

$$\int_2^8 \frac{1}{3x+5}\, dx$$

Write your answer in exact form or rounded to two decimal places.

7. Evaluate the definite integral below.

$$\int_1^2 (16x - 16)\left(4x^2 - 8x + 4\right)^3\, dx$$

Write your answer in exact form or rounded to two decimal places.

8. Find the average value of the function below on the interval that is given.

$$f(x) = \frac{4}{\sqrt{4x+1}}; [3,5]$$

Write your answer in exact form or rounded to two decimal places.

9. Evaluate the definite integral below.

$$\int_{-2}^1 \left(-3x\left(3x^2 + 4\right)^3\right)\, dx$$

Write your answer in exact form or rounded to two decimal places.

10. Evaluate the definite integral below.

$$\int_{-2}^{-1} 8xe^{\left(2x^2-1\right)}\, dx$$

Write your answer in exact form or rounded to two decimal places.

14.5 Area under a Curve (with Applications)

The Integral as Area

For $y = f(x)$, a continuous function on the interval $[a, b]$, the integral $\int_a^b f(x)dx$ represents:

1. The total area _____ from $x = a$ to $x = b$ if

$f(x)$ is _____ for all x in $[a, b]$.

2. The _____ above the x-axis and below the x-axis

that are bounded by the curve and the x-axis from $x = a$ to $x = b$ if $f(x)$ is _____ for

any x in $[a, b]$.

Example 1: Interpreting the Integral

a. Evaluate $\int_0^2 (x + 1)dx$ and interpret the integral geometrically.

Solution

Example 2: Total Area

Find the total area bounded by the x-axis and the curve $y = 4 - x^2$ on the interval $[-2, 3]$.

Solution

Additional Properties of the Definite Integral

$$\int_a^b f(x)dx =$$

_____

$$\int_a^b f(x)dx =$$ where c is any point with $a \leq c \leq b$.

_____

Example 3: Bounded Area

Find the area of the region bounded by the function

$$f(x) = \begin{cases} x+2 & \text{if } x \leq 1 \\ 5-2x & \text{if } x > 1 \end{cases}$$

and the x-axis from $x = -1$ to $x = 2$.

Solution

Example 4: Marginal Analysis

A picture frame maker knows that his profit (in dollars) is changing at a rate given by the function $P'(x) = 20 - 0.05x$, where x is the number of frames he makes and sells.

  a. Find the profit from selling the first 200 frames.

Solution

14.5 Exercises

1. Find the total area bounded by the x-axis and the curve $y = f(x)$ on the indicated interval. Write your answer in exact form or as a decimal number rounded to the nearest thousandth.

   $f(x) = 8x^3 - 5; [1, 4]$

2. Find the total area bounded by the x-axis and the curve $y = f(x)$ on the indicated interval. Write your answer in exact form or as a decimal number rounded to the nearest thousandth.

   $f(x) = 3x^2 + x + 3; [-2, 5]$

3. Find the total area bounded by the x-axis and the curve $y = f(x)$ on the indicated interval. Write your answer in exact form or as a decimal number rounded to the nearest thousandth.

   $f(x) = 5 + 8e^{0.4x}; [1, 5]$

4. Find the total area bounded by the x-axis and the curve $y = f(x)$ on the indicated interval. Write your answer in exact form or as a decimal number rounded to the nearest thousandth.

   $f(x) = \dfrac{1}{7x - 1}; [4, 5]$

5. The marginal profit for a certain style of sports jacket is given by $P'(x) = 60 - 0.57x$ dollars per jacket, where x is the number of sports jackets produced and sold weekly. Find the profit for the first 62 sports jackets that are produced and sold. Round your answer to the nearest cent.

6. The marginal profit of a product is given by $P'(x) = 30 - 0.081e^{0.1x}$ dollars per item, where x is the number of items produced and sold. Find the profit for the first 20 items. Round your answer to the nearest cent.

7. The marginal cost of a product is given by $158 + \dfrac{169}{\sqrt{x}}$ dollars per unit, where x is the number of units produced. The current level of production is 127 units weekly. If the level of production is increased to 207 units weekly, find the increase in the total costs. Round your answer to the nearest cent.

8. The marginal revenue from the sale of x bottles of wine is given by $6.6 - 0.12\sqrt{x}$ dollars per bottle. Find the increase in total revenue if the number of bottles sold is increased from 132 to 252. Round your answer to the nearest cent.

9. It was discovered that after t years a certain population of wild animals will increase at a rate of $P'(t) = 69 - 6t^{\frac{1}{3}}$ animals per year. Find the increase in the population during the first 10 years after the rate was discovered. Round your answer to the nearest whole animal.

10. Find the total area bounded by the x-axis and the curve $y = f(x)$ on the indicated interval.

$f(x) = x^2 + x - 2; [-5, 0]$

14.6 Area between Two Curves (with Applications)

Area Between Two Curves

If f and g are two continuous functions and $f(x) \geq g(x)$ on the interval $[a, b]$, then the area between the two curves on this interval is

_____

Example 1: Finding the Area Between Two Curves

Find the area enclosed by the curves $y = x^2 - 1$ and $y = 1 - x$.

Solution

Consumers' Surplus

The **consumers' surplus** is defined to be

$$CS = \underline{\hspace{5cm}},$$

where $p = D(x)$ is the _____ curve and (x_E, p_E) is the _____ .

Producers' Surplus

The **producers' surplus** is defined to be

$$PS = \underline{\hspace{5cm}},$$

where $p = S(x)$ is the _____ curve and (x_E, p_E) is the equilibrium point.

Example 3: Determining Surpluses

Suppose, for a certain new brand of car stereo, the demand function is $D(x) = 6 - x$ and the supply function is $S(x) = x^2$.

   a. Find the equilibrium point.

   b. Find the consumers' surplus.

   c. Find the producers' surplus.

Solution

Coefficient of Inequality for a Lorenz Curve

If $y = f(x)$ represents a Lorenz curve, then

$$\text{Coefficient of inequality} = \underline{\hspace{10cm}}.$$

Example 4: Using a Lorenz Curve

Suppose that $f(x) = 0.6x^2 + 0.4x$ represents a Lorenz curve for some country.

   a. What percent of the country's total income is earned by the lower 50 percent of the families in this country?

   b. Find the coefficient of inequality.

Solution

14.6 Exercises

1. The demand function for a particular product is given by the function $D(x) = -x^2 + 900$. Find the consumers' surplus if $x_E = 15$ units.

2. The supply function for a product is given by the function $S(x) = 5x^2 + 2304$. Find the producers' surplus if $x_E = 24$ units.

3. Consider the following demand and supply functions.

$$D(x) = 211 - 0.8x, \, S(x) = 3x + 116, 0 \le x \le 70$$

   a. Find the equilibrium point.

   b. Find the consumers' surplus at the equilibrium point. Round your answer to the nearest cent.

   c. Find the producers' surplus at the equilibrium point. Round your answer to the nearest cent.

4. The income distribution for country A is estimated by the function $f(x) = 0.7x - 0.63x^2 + 0.93x^3$. The income distribution for country B is estimated by the function $f(x) = 0.62x - 0.2x^2 + 0.58x^3$.

   a. Find the coefficient of inequality for each of the two countries. Round your answers to three decimal places.

   b. Which country has a more equitable income distribution?

5. Find the area of the region bounded by the graphs of the given equations. Write your answer in exact form or rounded to two decimal places.

$$y = e^{-0.4x}, y = 2 + x^2, x = 1, x = 3$$

6. Find the area of the region bounded by the graphs of the given equations.

$y = x^2 + 12x + 4, y = -x^2 + 4$

7. Find the area of the region bounded by the graphs of the given equations. Write your answer in exact form or rounded to two decimal places.

$y = e^{-x+1}, y = 4 + x, x = 1, x = 3$

14.7 Differential Equations

Solution of a Differential Equation

A function $y = f(x)$ is a _____ of a differential equation if the function and its

appropriate derivatives _____ .

Example 1: Solutions of Differential Equations

Verify that the function $y = cx^3 - 4$ is a general solution of the differential equation $xy' - 3y = 12$.

Solution

To Solve a First-Order Separable Differential Equation

Step 1: Write y' as $\dfrac{dy}{dx}$.

Step 2: Separate variables by writing _____ on

one side of the equation and _____ on the other

side.

Step 3: _____ of the new equation.

Example 3: Using the Method of Separating Variables

Solve the differential equation $y' = \dfrac{y}{x - 1}$ where $x > 1$ and $y > 0$.

Solution

Example 5: Initial-Value Problem

Solve the initial-value problem $y' = x$ with initial condition $f(0) = 1$. (That is, $y = 1$ when $x = 0$.)

Solution

Example 7: Population of Bacteria

At the beginning of an experiment, a culture of 100 bacteria is growing in a medium that will allow a maximum of 10,000 bacteria to survive. If the population is 200 after 5 hours, use the logistic equation to represent the number of bacteria present t hours after the experiment has begun.

Solution

14.7 Exercises

1. Solve the differential equation given below.

$$\frac{dy}{dx} = 4x^3 y$$

2. Solve the differential equation given below.

$$\frac{dy}{dx} = \frac{7y + 2}{5x + 8}$$

3. Solve the differential equation given below.

$$\frac{dy}{dx} = -4y$$

4. Solve the differential equation given below.

$$\frac{dy}{dx} = -6\left(8 - y\right)$$

5. Solve the differential equation given below.

$$\frac{dy}{dx} = 4y\left(2 - y\right)$$

6. Solve the initial value problem given below.

$$\frac{dy}{dx} = -12x^3 y \text{ and } y = 3 \text{ when } x = -1$$

CHAPTER 15

Additional Integration Topics

15.1 Integration by Parts

Formula for Integration by Parts

$$\int u \cdot dv = \underline{\hspace{4cm}}$$

Example 2: Integration by Parts

Find $\displaystyle\int x \ln x \, dx$.

Solution

Example 3: Integration by Parts

Find $\displaystyle\int x^2 e^{-x} \, dx$.

Solution

15.1 Exercises

1. Use integration by parts to solve the integral below.

$$\int \ln(x) \cdot x^4 \, dx$$

2. Use integration by parts to solve the integral below.

$$\int \ln(x) \cdot x \, dx$$

3. Use integration by parts to solve the integral below.

$$\int \ln(4x) \cdot x^7 \, dx$$

4. Use integration by parts to solve the integral below.

$$\int \ln(3x) \cdot x^4 \, dx$$

5. Use integration by parts to solve the integral below.

$$\int x(x-9)^{-4} \, dx$$

6. Use integration by parts to solve the integral below.

$$\int x\sqrt{x-2} \, dx$$

7. Use integration by parts to solve the integral below.

$$\int -9xe^{2x}\ dx$$

8. Use integration by parts to solve the integral below.

$$\int xe^x\ dx$$

9. Use integration by parts to evaluate the definite integral below.

$$\int_{-5}^{4} x\sqrt{x+5}\ dx$$

10. Use integration by parts to evaluate the definite integral below.

$$\int_{-1}^{0} x\sqrt{x+1}\ dx$$

15.2 Annuities and Income Streams

Future Value of an Annuity

The **amount** (or **future value**) **of an annuity** at the end of N time periods is approximated by the integral

$$\int_0^N Pe^{r(N-t)}dt = \underline{\hspace{4cm}},$$

where P is the number of dollars invested each time period, and r is the interest rate (as a decimal) per time period.

Example 1: Future Value of an Annuity

Suppose that the parents of a child set up an annuity account paying 10 percent compounded continuously for the child's college education. They deposit $500 each year for 20 years. What will be the approximate amount of the annuity in 20 years?

Solution

Future Value of an Income Stream

The **amount** (or **future value**) **of an income stream** at the end of T years is given by the integral

$$A = \int_0^T R(t) \underline{\hspace{3cm}} dt,$$

where $R(t)$ is the rate of flow of revenue at time t, r is the annual interest rate (as a decimal), and interest is compounded continuously.

Example 2: Future Value of an Income Stream

A travel agency expects income from airline ticket sales to increase at a continuous rate represented by $R(t) = 210 + 0.1t$ in hundreds of dollars per year over the next 3 years. With interest at 5 percent compounded continuously, what income does the agency expect from airline ticket sales over the next 3 years?

Solution

Present Value of an Income Stream

The **present value of an income stream** over T years is given by the integral

$$PV = \int_0^T R(t) \underline{\hspace{3cm}} dt,$$

where $R(t)$ is the rate of flow of revenue at time t, r is the annual interest rate (as a decimal), and interest is compounded continuously.

Example 3: Present Value of an Income Stream

A new department store is expected to generate income at a continuous rate described by the function $R(t) = 50{,}000 + 2000t$ dollars per year over the next 5 years. Find the present value of the store if the current interest rate is 10 percent compounded continuously.

Solution

15.2 Exercises

1. Tim has decided to invest $1500 each year into an annuity. If the account pays interest at a rate of 5 percent compounded continuously, find Tim's account balance after 9 years. Round intermediate answers to eight decimal places and final answer to two decimal places.

2. Keisha has decided to invest $8500 each year into an annuity. If the account pays interest at a rate of 4 percent compounded continuously, find Keisha's account balance after 20 years. Round intermediate answers to eight decimal places and final answer to two decimal places.

3. Find the value of an income stream after 6 years if the rate of flow is estimated to be $280,000 annually and the income is deposited at a rate of 8 percent compounded continuously. Round intermediate answers to eight decimal places and final answer to two decimal places.

4. Find the value of an income stream after 17 years if the rate of flow is estimated to be $180,000 annually and the income is deposited at a rate of 9 percent compounded continuously. Round intermediate answers to eight decimal places and final answer to two decimal places.

5. Find the value of an income stream after 20 years if $R(t) = 4500e^{-0.04t}$ is the rate of flow of revenue and the income is deposited at a rate of 5 percent compounded continuously. Round intermediate answers to eight decimal places and final answer to two decimal places.

6. Find the value of an income stream after 9 years if $R(t) = 8000e^{0.04t}$ is the rate of flow of revenue and the income is deposited at a rate of 8 percent compounded continuously. Round intermediate answers to eight decimal places and final answer to two decimal places.

7. Find the present value of an income stream with $R(t) = 330 + 0.8t$, $r = 5$ percent, and $T = 11$. Round intermediate answers to eight decimal places and final answer to two decimal places.

8. Find the present value of an income stream with $R(t) = 370 + 0.5t$, $r = 5$ percent, and $T = 11$. Round intermediate answers to eight decimal places and final answer to two decimal places.

9. The rate of flow of an income stream for the next 5 years is estimated by $R(t) = 1000e^{-0.03t}$. Find the present value of this flow if the interest rate is 7 percent compounded continuously. Round intermediate answers to eight decimal places and final answer to two decimal places.

10. The rate of flow of an income stream for the next 4 years is estimated by $R(t) = 7000e^{-0.05t}$. Find the present value of this flow if the interest rate is 5 percent compounded continuously. Round intermediate answers to eight decimal places and final answer to two decimal places.

15.3 Tables of Integrals

Example 2: Using the Integral Table

Evaluate $\int \dfrac{9}{x^2 - 25}\,dx$.

Solution

Example 3: Using a Reduction Formula

Integrate $\int x^2 e^{-3x}\,dx$.

Solution

Example 4: Rewriting the Integrand

Integrate $\int \dfrac{x+1}{x^2 + 5x + 6}\,dx$.

Solution

15.3 Exercises

1. Use an integral table to evaluate the integral below.

$$\int 9x^5 \ln(x) \ dx$$

2. Use an integral table to evaluate the integral below.

$$\int -8x^4 \ln(x) \ dx$$

3. Use an integral table to evaluate the integral below.

$$\int \frac{1}{\sqrt{x^2 - 144}} \ dx$$

4. Use an integral table to evaluate the integral below.

$$\int \frac{1}{\sqrt{x^2 - 11}} \ dx$$

5. Use an integral table to evaluate the integral below.

$$\int \frac{1}{x^2 + 14x + 45} \ dx$$

6. Use an integral table to evaluate the integral below.

$$\int \frac{1}{5x^2 + 9x + 4} \ dx$$

7. Use an integral table to evaluate the integral below.

$$\int \frac{5}{x\,(2x-17)}\,dx$$

8. Use an integral table to evaluate the integral below.

$$\int \frac{-5}{x\,(3x-4)}\,dx$$

9. Use an integral table to evaluate the integral below.

$$\int x^2 e^{-2x}\,dx$$

10. Use an integral table to evaluate the integral below.

$$\int x^2 e^{3x}\,dx$$

15.4 Numerical Integration

Trapezoidal Rule

An approximation of $\int_a^b f(x)dx$ using the Trapezoidal Rule is the _____

$$T_n = \frac{\Delta x}{2}\left[f(x_0) + 2f(x_1) + 2f(x_2) + \cdots + 2f(x_{n-1}) + f(x_n)\right],$$

where $\Delta x = \dfrac{(b-a)}{n}$ and $x_i = a + i\Delta x$ for $i = 0, 1, \ldots, n$.

Example 1

Use the Trapezoidal Rule to approximate $\int_1^2 \ln x\, dx$ with $n = 10$, and compare the result to the exact value of this integral.

Solution

Error Estimate for the Trapezoidal Rule

If f'' is continuous on $[a, b]$ and M is an upper bound for $|f''(x)|$, then the error E_T between the exact

value of $\int_a^b f(x)dx$ and the Trapezoidal Rule approximation T_n satisfies

$$|E_T| \leq \underline{\hspace{3cm}}$$

where n is the number of subintervals in the partition.

Simpson's Rule

An approximation of $\int_a^b f(x)dx$ using Simpson's Rule is the sum

$$S_n = \frac{\Delta x}{3}\left[\begin{array}{l} f(x_0) + 4f(x_1) + 2f(x_2) + 4f(x_3) + 2f(x_4) + \cdots \\ +2f(x_{n-2}) + 4f(x_{n-1}) + f(x_n) \end{array}\right]$$

where n is even, $\Delta x = \dfrac{(b-a)}{n}$, and $x_i = a + i\Delta x$ for $i = 0, 1, \ldots, n$. (Note the pattern of the

coefficients: _____ .)

Example 4

Use Simpson's Rule to approximate $\int_1^2 \ln x\, dx$ with $n = 10$.

Solution

Error Estimate for Simpson's Rule

If $f^{(4)}$ is continuous on $[a, b]$ and M is an upper bound for $\left|f^{(4)}(x)\right|$, then the error E_S between the exact

value of $\int_a^b f(x)dx$ and the Simpson's Rule approximation S_n satisfies

$$|E_S| \leq \underline{\hspace{3cm}}$$

where n is the number of subintervals in the partition.

Example 5

Use Simpson's Rule with $n = 4$ to approximate the nonelementary integral $\int_0^1 e^{x^2}\, dx$, and estimate the error in

the approximation.

Solution

Example 6

A landscape designer has planned a free-form garden pond with the shape shown in Figure 6 and needs to estimate its volume in cubic feet. The pond will have a uniform depth of 2 feet. At 1-foot intervals, the distances across the pond are to be as indicated in the diagram. Use Simpson's Rule to estimate the volume of the pond.

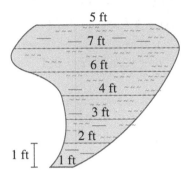

Figure 6: Garden Pond

Solution

15.4 Exercises

1. Use the Trapezoidal Rule with $n = 8$ to approximate the integral $\int_0^4 32x^4 dx$. Round any intermediate calculations, if needed, to no less than six decimal places, and round your final answer to four decimal places.

2. Use Simpson's Rule with $n = 8$ to approximate the integral $\int_0^8 11x^4 dx$. Round any intermediate calculations, if needed, to no less than six decimal places, and round your final answer to four decimal places.

3. Use the Trapezoidal Rule with $n = 8$ to approximate the integral $\int_1^5 \frac{6}{x} dx$. Round any intermediate calculations, if needed, to no less than six decimal places, and round your final answer to four decimal places.

4. Use Simpson's Rule with $n = 8$ to approximate the integral $\int_1^5 \frac{6}{x} dx$. Round any intermediate calculations, if needed, to no less than six decimal places, and round your final answer to four decimal places.

5. Use the Trapezoidal Rule with $n = 10$ to approximate the integral $\int_0^{10} \sqrt[3]{2x} dx$. Round any intermediate calculations, if needed, to no less than six decimal places, and round your final answer to four decimal places.

6. Use Simpson's Rule with $n = 10$ to approximate the integral $\int_0^{10} \sqrt[3]{2x} dx$. Round any intermediate calculations, if needed, to no less than six decimal places, and round your final answer to four decimal places.

7. Use the Trapezoidal Rule to estimate the amount of water needed to raise the water level by two inches in a free-form pool. Measured at 2-foot intervals from one edge of the pool to the other edge, the distances across the pool (in feet) are the following: 4, 10, 13, 15, 16, 15, 13, 12, 11, 10, 9, 8. Write the exact answer. Do not round.

8. Measured at 5-foot intervals from one bank of the Lazee river, where the Dinkatown ferry docks, to the other bank, the depths of the river (in feet) are the following: 0, 4, 13, 20, 25, 27, 27, 24, 17, 4, 0. If the river flows at 5 ft/s, use Simpson's Rule to estimate the amount of water passing by the river dock every second. Round your answer to the nearest whole number if necessary.

15.5 Improper Integrals

Improper Integral

The integral $\int_a^{+\infty} f(x)\, dx$ is called an **improper integral**. This integral is defined as the following limit:

$$\int_a^{+\infty} f(x)\, dx = \lim_{b \to +\infty} \int_a^b f(x)\, dx.$$

If $\lim\limits_{b \to +\infty} \int_a^b f(x)\, dx$ exists, then the improper integral is said to be _____.

If $\lim\limits_{b \to +\infty} \int_a^b f(x)\, dx$ does not exist, then the improper integral is said to be _____.

Example 1: Evaluating an Improper Integral

Evaluate $\int_1^{+\infty} \dfrac{1}{x^2}\, dx$.

Solution

Example 3: Determining Integral Convergence

Determine whether the improper integral $\int_1^{+\infty} \dfrac{1}{x}\, dx$ is convergent or divergent. Evaluate it if it is convergent.

Solution

Example 6: Area Bounded by an Asymptote

Find the value of $\displaystyle\int_0^1 \frac{1}{x^2}\, dx$.

Solution

15.5 Exercises

1. Find the limit.

$$\lim_{b\to+\infty} \frac{3}{5\sqrt[3]{b}}$$

2. Determine whether the improper integral is convergent or divergent. If the improper integral is convergent, evaluate.

$$\int_6^{+\infty} \frac{2}{15x^2}\, dx$$

3. Determine whether the improper integral is convergent or divergent. If the improper integral is convergent, evaluate.

$$\int_4^{+\infty} 3e^{-2x}\, dx$$

4. Find the area, if it exists, of the region under the curve on the given interval of the x-axis.

$$f(x) = \frac{2}{7x^3}, x \geq 5$$

5. Determine whether the improper integral is convergent or divergent. If the improper integral is convergent, evaluate.

$$\int_4^{+\infty} \frac{9}{x^{\frac{1}{3}}}\, dx$$

6. Determine whether the improper integral is convergent or divergent. If the improper integral is convergent, evaluate.

$$\int_3^{+\infty} \frac{3}{x^3}\, dx$$

7. Determine whether the improper integral is convergent or divergent. If the improper integral is convergent, evaluate.

$$\int_2^{+\infty} 2e^{-2x}\, dx$$

8. Determine whether the improper integral is convergent or divergent. If the improper integral is convergent, evaluate.

$$\int_3^{+\infty} \frac{2}{\sqrt{(8x+3)}}\, dx$$

9. Determine whether the improper integral is convergent or divergent. If the improper integral is convergent, evaluate.

$$\int_3^{+\infty} 4xe^{-x^2}\, dx$$

10. Find the area, if it exists, of the region between the curve and the x-axis on the given interval.

$$f(x) = 24e^{-0.8x}, x \geq 2$$

15.6 Volume

Volume of a Solid of Revolution

If $y = f(x)$ is a nonnegative continuous function on the interval $[a, b]$, then the volume of the solid formed by revolving the region bounded by the graph of the function and the x-axis ($a \leq x \leq b$) about the x-axis is given by

$$V = \underline{\hspace{4cm}} .$$

Example 1: Finding the Volume of a Solid of Revolution

Find the volume of the solid of revolution generated by revolving the region under the curve $y = \sqrt{x}$ from $x = 0$ to $x = 4$ about the axis.

Solution

Example 2: Finding the Volume of a Solid of Revolution

If the region under the line $y = \dfrac{r}{h} x$ from $x = 0$ to $x = h$ is revolved about the x-axis, a circular cone is formed. Find the volume of this cone.

Solution

15.6 Exercises

1. Consider the function $y = 3x + 3$ between $x = 4$ and $x = 6$. Find the volume of the solid generated when the region bounded by the graphs of the given equations and the x-axis is rotated about the x-axis.

2. Consider the function $y = 64 - x^2$ between $x = 2$ and $x = 8$. Find the volume of the solid generated when the region bounded by the graphs of the given equations and the x-axis is rotated about the x-axis.

3. Consider the function $y = 2\sqrt{x}$ between $x = 4$ and $x = 9$. Find the volume of the solid generated when the region bounded by the graphs of the given equations and the x-axis is rotated about the x-axis.

4. Consider the function $y = 5x^2$ between $x = 0$ and $x = 7$. Find the volume of the solid generated when the region bounded by the graphs of the given equations and the x-axis is rotated about the x-axis.

5. Consider the function $y = e^{2x}$ between $x = 1$ and $x = 5$. Find the volume of the solid generated when the region bounded by the graphs of the given equations and the x-axis is rotated about the x-axis.

6. Consider the function $y = \sqrt{1 - x^2}$ between $x = -1$ and $x = 1$. Find the volume of the solid generated when the region bounded by the graphs of the given equations and the x-axis is rotated about the x-axis.

7. Consider the function $y = \dfrac{4}{x}$ between $x = 2$ and $x = 7$. Find the volume of the solid generated when the region bounded by the graphs of the given equations and the x-axis is rotated about the x-axis.

8. Consider the function $y = \dfrac{1}{x^2}$ between $x = 2$ and $x = 4$. Find the volume of the solid generated when the region bounded by the graphs of the given equations and the x-axis is rotated about the x-axis.

9. Consider the function $y = \dfrac{3}{\sqrt{x}}$ between $x = 2$ and $x = 3$. Find the volume of the solid generated when the region bounded by the graphs of the given equations and the x-axis is rotated about the x-axis.

10. Consider the function $y = 8x + 2\sqrt{x}$ between $x = 1$ and $x = 4$. Find the volume of the solid generated when the region bounded by the graphs of the given equations and the x-axis is rotated about the x-axis.

CHAPTER 16

Multivariable Calculus

16.1 Functions of Several Variables

Domain and Range of $f(x, y)$

If D is a set of ordered pairs of real numbers and for each ordered pair (x, y) in D there corresponds a unique real number $f(x, y)$, then f is called a **function of x and y**. The set D is the _____ of the function. The set of all the values of $f(x, y)$ is called the _____ of the function.

Example 1: Evaluating $f(x, y)$

For $f(x, y) = 3x + y^2$ find the following.

a. $f(2, 3)$

b. $f\left(2, \sqrt{2}\right)$

c. $f(0, 0)$.

Solution

Example 2: Evaluating $f(x, y)$

For $f(x, y) = e^{\sqrt{x}} + \ln y$,

a. find the domain and

b. find $f(0, 1)$.

Solution

Example 3: Evaluating Revenue Influenced by Two Variables

A pharmacy sells two brands of aspirin. Brand A sells for $1.25 per bottle and Brand B sells for $1.50 per bottle.

  a. What is the revenue function for aspirin?

  b. What is the revenue for aspirin if 100 bottles of Brand A and 150 bottles of Brand B are sold?

Solution

Example 5: Using the Cobb-Douglas Production

Suppose that the function $P(x, y) = 500x^{0.3}y^{0.7}$ represents the number of units produced by a company with x units of labor and y units of capital.

  a. How many units of a product will be manufactured if 300 units of labor and 50 units of capital are used?

  b. How many units will be produced if twice the number of units of labor and capital are used?

Solution

16.1 Exercises

  1. Calculate the value of $f(-2, 10)$ for the given function. Write your answer as an integer or simplified fraction.

  $$f(x, y) = -9x - 10xy - 11y$$

2. Calculate the value of $f(5, -8)$ for the given function. Write your answer as an integer or simplified fraction.

$$f(x, y) = \frac{8x + 15y}{-3x + 10y}$$

3. Calculate the value of $f(3, 2)$ for the given function. Write your answer as an integer or simplified fraction.

$$f(x, y) = \frac{-11xy^2}{\sqrt{12x + y^2}}$$

4. Calculate the value of $f(-9, 5, 4)$ for the given function. Write your answer as an integer or simplified fraction.

$$f(x, y, z) = 7xy - 2xz + 9yz$$

5. The number of units of a product that are manufactured by a company is given by $f(L, K) = 300L^{0.2}K^{0.8}$, where L is the units of labor and K is the units of capital.

    a. How many units of a product will be manufactured by utilizing 59 units of labor and 22 units of capital? Round your answer to the nearest unit.

    b. How many units of a product will be manufactured if the number of units of labor and capital given in Part 1 are doubled? Round your answer to the nearest unit.

6. A company manufactures two models of a product, model A and model B. The cost for model A is \$22 and the cost for model B is \$33. If the fixed costs are \$650, the total cost function is given by $C(x, y) = 650 + 22x + 33y$, where x is the number of model A and y is the number of model B. Find $C(23, 55)$.

7. A company makes two grades of paint, grade I, guaranteed for 5 years, and grade II, guaranteed for 10 years. A gallon of grade I costs \$3.20 to make, while a gallon of grade II costs \$3.50 to make. The weekly fixed costs are \$5000.

    a. Find the cost function $C(x, y)$ for making x gallons of grade I and y gallons of grade II.

    b. What is the cost of making 110 gallons of grade I and 130 gallons of grade II?

8. A grocery store sells two brands of a product, the "house" brand and a "name" brand. The manager estimates that if she sells the "house" brand for x dollars and the "name" brand for y dollars, she will be able to sell $70 - 20x + 17y$ units of the "house" brand and $104 + 17x - 23y$ units of the "name" brand.

    a. Find the revenue function $R(x, y)$.

    b. What is the revenue if she sells the "house" brand for \$4.30 and the "name" brand for \$4.40?

16.2 Partial Derivatives

Partial Derivatives

Let $z = f(x, y)$

    a. The **first partial derivative of f with respect to x** (if it exists) is

$$\frac{\partial f}{\partial x} = \underline{\hspace{8cm}}$$

    b. The **first partial derivative of f with respect to y** (if it exists) is

$$\frac{\partial f}{\partial y} = \underline{\hspace{8cm}}$$

    c. $\dfrac{\partial f}{\partial x} = \dfrac{dz}{dx}$ and $\dfrac{\partial f}{\partial y} = \dfrac{dz}{dy}$ in our notation.

Example 2: Finding a Partial Derivative

For the function $f(x, y) = e^{xy} - \ln x + y^3$, find the following.

    a. f_x

    b. f_y

Solution

Example 3: Finding a Partial Derivative

For the function $f(x, y) = xe^{xy} + y^2$, find the following.

    a. $f_x(1, 2)$

    b. $f_y(1, 2)$

Solution

Example 4: Using the Partial Derivative

Given the function $z = x^2 + xy + y^2 - 7x - 8y + 1$, find all the ordered pairs (x, y) where both $\dfrac{\partial z}{\partial x} = 0$ and $\dfrac{\partial z}{\partial y} = 0$.

Solution

Example 5: Using the Cobb-Douglas Production Formula

Suppose that the production function $P(x, y) = 2000x^{0.5}y^{0.5}$ is known. Determine the marginal productivity of labor and the marginal productivity of capital when 16 units of labor and 144 units of capital are used.

Solution

Example 7: Finding Partial Derivatives with 3 Variables

For $w = 2ye^{xy} + z^2$, find the following.

a. $\dfrac{\partial w}{\partial x}$

b. $\dfrac{\partial w}{\partial y}$

c. $\dfrac{\partial w}{\partial z}$

Solution

16.2 Exercises

1. Consider the following function:

$$f(x, y) = y^3 \sqrt{16 + x^2}$$

    a. Find $\dfrac{\partial f}{\partial x}$.

    b. Find $\dfrac{\partial f}{\partial y}$.

2. Consider the following function:

$$f(x, y) = \ln\left(-5x^4 + 4y^5\right)$$

    a. Find f_x.

    b. Find f_y.

3. Consider the following function:

$$f(x, y) = y^2 \ln\left(-4x^5 + 5y^3\right)$$

    a. Find f_x.

b. Find f_y.

4. Consider the following function:

$$f(x, y) = -4x^2 e^y + 3ye^{x^3}$$

a. Find f_x.

b. Find f_y.

16.3 Local Extrema for Functions of Two Variables

Local Extrema for a Function $z = f(x, y)$

Suppose that $z = f(x, y)$ is a function defined on a region in the xy-plane and (a, b) is a point in that region.

1. If there is an open region R containing (a, b) such that

$$f(a, b) \geq f(x, y)$$

for all (x, y) in R, then $f(a, b)$ is called a _____ of f.

2. If there is an open region R containing (a, b) such that

$$f(a, b) \leq f(x, y)$$

for all (x, y) in R, then $f(a, b)$ is called a _____ of f.

Example 1: Locating Critical Points

Locate the critical points for the function $f(x, y) = 4 + (x - 3)^2 + (y + 5)^2$.

Solution

Second Partials Test (or D-Test)

Suppose that a function $z = f(x, y)$ and the first partial derivatives and the second partial derivatives are all defined in an open region R and that (a, b) **is a critical point** in R such that

$$f_x(a, b) = 0 \text{ and } f_y(a, b) = 0$$

Define the quantity D as follows:

$$D = f_{xx}(a, b) \cdot f_{yy}(a, b) - \left[f_{xy}(a, b)\right]^2.$$

Case 1: If $D > 0$ and $f_{xx}(a, b) > 0$, then $f(a, b)$ is a _____ .

Case 2: If $D > 0$ and $f_{xx}(a, b) < 0$, then $f(a, b)$ is a _____ .

Case 3: If $D < 0$, then $(a, b, f(a, b))$ is a _____ .

Case 4: If $D = 0$, then this test gives _____ .

Example 2: Applying the D-test

Find all the local minima, local maxima, and saddle points for the function

$$f(x, y) = x^2 - xy + y^2 - 9x + 5.$$

Solution

Example 3: Locating and Classifying Critical Points

Locate and classify all critical points of the function

$$g(x, y) = \frac{1}{3}x^3 - \frac{1}{3}y^3 + xy.$$

Solution

Example 4: Optimizing an Equation in Two Variables

A company produces and sells two styles of umbrellas. One style sells for \$20 each and the other sells for \$25 each. The company has determined that if x thousand of the first style and y thousand of the second style are produced, then the total cost in thousands of dollars is given by the function

$$C\left(x, y\right) = 3x^2 - 3xy + \frac{3}{2}y^2 + 32x - 29y + 70.$$

How many of each style of umbrella should the company produce and sell in order to maximize profit?

Solution

16.3 Exercises

1. Find all local maxima, local minima, and saddle points for the function given below. Write your answer in the form $\left(x, y, z\right)$.

$$f\left(x, y\right) = -6x^2 + 5y^2 + 4x + 5y - 8$$

Separate multiple points with a comma.

2. Find all local maxima, local minima, and saddle points for the function given below. Write your answer in the form $\left(x, y, z\right)$.

$$f\left(x, y\right) = -2x + 5y + 8x^2 + y^2 + 7$$

Separate multiple points with a comma.

3. Find all local maxima, local minima, and saddle points for the function given below. Write your answer in the form (x, y, z). Separate multiple points with a comma.

$$f(x, y) = 2x^3 + y^2 - 24x - 2y + 16$$

4. A manufacturer makes two kinds of magazine racks, metal and plastic. The metal magazine racks sell for $52 and the plastic magazine racks for $31. The total cost function is $C(x, y) = 0.25x^2 + 0.5y^2 - 28x - 33y + 2520$ dollars, where x is the number of metal magazine racks and y is the number of plastic magazine racks. How many magazine racks of each kind should be produced and sold to maximize the profit?

5. Find all local maxima, local minima, and saddle points for the function given below. Write your answer in the form (x, y, z). Separate multiple points with a comma.

$$f(x, y) = -3x^2 + y^3 + 3y^2 + 12x - 9y + 9$$

6. Find all local maxima, local minima, and saddle points for the function given below. Write your answer in the form (x, y, z). Separate multiple points with a comma.

$$f(x, y) = -2x^2 - x^2 y - 4y^2$$

7. Find all local maxima, local minima, and saddle points for the function given below. Write your answer in the form (x, y, z). Separate multiple points with a comma.

$$f(x, y) = 2xy + \frac{8}{x} - \frac{4}{y}$$

16.4 Lagrange Multipliers

The Method of Lagrange Multipliers

To maximize or minimize a function $z = f(x, y)$ subject to the constraint $g(x, y) = 0$:

Step 1: Form the Lagrange function

_____

Step 2: Find each of the partial derivatives F_x, F_y, and F_λ, provided all partials exist.

Step 3: Solve the system of three equations

$$\begin{cases} F_x(x, y, \lambda) = 0 \\ F_y(x, y, \lambda) = 0 \\ F_\lambda(x, y, \lambda) = 0. \end{cases}$$

Step 4: If f has a local maximum or local minimum subject to the constraint $g(x, y) = 0$, then the corresponding x and y values will be found among the solutions to the system in Step 3.

Example 2: Applying the Method of Lagrange Multipliers

Use the method of Lagrange multipliers to find the maximum value of the function $f(x, y) = xy$ under the restriction $g(x, y) = x^2 + y^2 - 200 = 0$, where $x > 0$ and $y > 0$.

Solution

Example 3: Optimization Using the Cobb–Douglas Production Formula

Suppose that the Cobb-Douglas Production Formula for a particular product is $f(x, y) = 100x^{0.6}y^{0.4}$, where x represents units of labor at a cost of \$200 per unit and y represents units of capital at a cost of \$300 per unit. If the company's budget allows \$120,000 for labor and capital, then the constraint is represented by the equation $200x + 300y = 120,000$. Find the maximum level of production for this product.

Solution

16.4 Exercises

1. Use the method of Lagrange multipliers to find the minimum value of f subject to the given constraint.

$f(x, y) = x^2 + y^2 - xy$, subject to $4x + 2y = -56$

2. Use the method of Lagrange multipliers to find the maximum value of f subject to the given constraint.

$f(x, y) = -2x^2 - 2y^2 + xy$, subject to $2x + 3y = 64$

3. A company has a plant in Phoenix and a plant in Charleston. The firm is committed to produce a total of 228 units of a product each week. The total weekly cost is given by $C(x, y) = \dfrac{8}{7}x^2 + \dfrac{1}{7}y^2 + 21x + 33y + 450$, where x is the number of units produced in Phoenix and y is the number of units produced in Charleston. How many units should be produced in each plant to minimize the total weekly cost?

4. A furniture store sells two styles of a chair, reclining and non-reclining. During the month of March, the management expects to sell exactly 477 chairs. The monthly profit is given by

$P(x, y) = -\dfrac{1}{3}x^2 - 3y^2 - \dfrac{1}{3}xy + 10x + 7y - 400$, where x is the number of reclining chairs sold and y is the number of non-reclining chairs sold. How many of each type should be sold to maximize the profit in March?

5. The marketing manager of a department store has determined that revenue, in dollars, is related to the number of units of television advertising, x, and the number of units of newspaper advertising, y, by the function $R(x, y) = 400 \left(182x - 2y^2 + 2xy - 4x^2\right)$. Each unit of television advertising costs \$3600, and each unit of newspaper advertising costs \$900. If the amount spent on advertising is \$65700, find the maximum revenue.

16.5 The Method of Least Squares

The Least-Squares Regression Line

For a set of data points $(x_1, y_1), (x_2, y_2), \ldots, (x_n, y_n)$, the regression line is _____
where (omitting indices in the summations):

1. $m = \dfrac{n \cdot \sum (xy) - \sum x \sum y}{n \cdot \sum x^2 - \left(\sum x\right)^2}$,

2. $b = \overline{y} - m\overline{x}$,

   and

3. $\overline{y} = \dfrac{1}{n} \sum y$ and $\overline{x} = \dfrac{1}{n} \sum x$.

Example 1: Using Linear Regression on Tabular Data

Use the formulas to find the regression line for the data given in the following table.

Solution

	x	y	xy	x^2
	4	3	12	16
	5	5	25	25
	12	5	60	144
	16	8	128	256
	18	7	126	324
Sums	55	28	351	765

16.5 Exercises

1. Consider the following points.

$(4, 5), (7, 10), (10, 14), (13, 17), (16, 18)$

 a. Find the equation of the regression line for the given points. Round any final values to the nearest hundredth, if necessary.

 b. Draw the scatter diagram and graph the regression line.

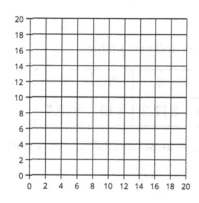

2. Consider the following points.

$(2, 16), (4, 7), (6, 7), (8, 7), (10, 1)$

 a. Find the equation of the regression line for the given points. Round any final values to the nearest hundredth, if necessary.

 b. Draw the scatter diagram and graph the regression line.

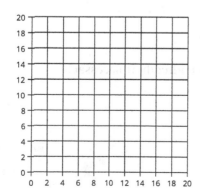

3. Find the equation of the regression line for the given points. Round any final values to the nearest hundredth, if necessary.

$(3.2, 16.1)$, $(1, 14.9)$, $(-1.2, 15.4)$, $(-3.4, 17.1)$, $(-5.6, 17.8)$

4. Find the equation of the regression line for the given points. Round any final values to the nearest hundredth, if necessary.

$(5.9, 15.2)$, $(13.1, 23.3)$, $(20.3, 31.2)$, $(27.5, 37)$, $(34.7, 44.5)$

5. The annual revenue (in millions of dollars) for a corporation is given in the following table.

Year (x)	1	2	3	4	5	6
Revenue (in millions) (y)	$43.6	$44.1	$46.9	$47.2	$49	$49.7

a. Find the line of regression for the data. Round any final values to the nearest hundredth, if necessary.

b. Estimate the revenue (in millions) for Year 7. Round to the nearest hundredth, if necessary.

6. The annual revenue (in millions of dollars) for a corporation is given in the following table.

Year (x)	1	2	3	4	5	6
Revenue (in millions) (y)	$39.5	$40.4	$38.6	$38.1	$38.9	$38.1

a. Find the line of regression for the data. Round any final values to the nearest hundredth, if necessary.

b. Estimate the revenue (in millions) for Year 7. Round to the nearest hundredth, if necessary.

16.6 Double Integrals

Definition of Double Integral

For a function $z = f(x, y)$ continuous on a region R in the xy-plane

$$\iint\limits_R f(x, y)\ dA = \iint\limits_R f(x, y)\ dx\ dy = \underline{\hspace{6cm}}$$

Value of the Double Integral

If $z = f(x, y)$ is a continuous function on a region R in the xy-plane and $f(x, y) \geq 0$ on R, then the volume of the solid bounded above by the graph of f and below by R is the **value of the double integral** of f on R:

$$V = \underline{\hspace{5cm}}$$

Theorem for Evaluating Double Integrals

If $z = f(x, y)$ is a continuous function on a closed region R, and

Case 1: R is of _____ **, then**

$$\iint\limits_R f(x, y)\,dx\,dy = \int_c^d \left(\int_a^b f(x, y)dx \right) dy = \int_a^b \left(\int_c^d f(x, y)dy \right) dx$$

Case 2: R is of _____ **, then**

$$\iint\limits_R f(x, y)\,dx\,dy = \int_a^b \left(\int_{g_1(x)}^{g_2(x)} f(x, y)dy \right) dx$$

Case 3: R is of _____ **, then**

$$\iint\limits_R f(x, y)\,dx\,dy = \int_c^d \left(\int_{h_1(y)}^{h_2(y)} f(x, y)dx \right) dy$$

Note: Only in Case 1 have we shown that the order of integration may be reversed. There are problems of the types in Case 2 and Case 3 in which the order of integration may be reversed; however, we will not discuss reversing the order of integration for such problems in this text.

Example 1: Evaluating a Double Integral on a Type I Region

Evaluate the double integral $\int_1^3 \int_1^e \frac{y^2}{x} dx dy$.

Solution

Example 2: Evaluating a Double Integral on a Type II Region

Evaluate the double integral $\int_0^1 \int_{x^2}^{x+1} 2x^2 y\, dx dy$.

Solution

Example 3: Evaluating a Double Integral on a Type III Region

Evaluate the double integral $\int_0^1 \int_0^{\frac{y}{3}} e^{3x+y} dx \, dy$.

Solution

16.6 Exercises

1. Evaluate the double integral.

$$\int_{-4}^{-2} \int_{-5}^{-3} (x+8) \, dy \, dx$$

2. Evaluate the double integral.

$$\int_{-3}^{-1} \int_{1}^{2} (11x - 3y) \, dx \, dy$$

3. Evaluate the double integral.

$$\int_{2}^{4} \int_{2}^{5} \left(\frac{8y}{3x} \right) dy \, dx$$

4. Evaluate the double integral on the given region.

$$\iint_R \left(y\sqrt{x+2} \right) dA; R : 0 \leq x \leq 2 \text{ and } 2 \leq y \leq 4$$

5. Evaluate the double integral on the given region.

$$\iint\limits_R \left(e^{x+6y}\right) \, dA; R : 2 \le x \le 4 \text{ and } 1 \le y \le 3$$

6. Evaluate the double integral.

$$\int_0^3 \int_{-8}^{1-y^2} \left(y\sqrt{x+8}\right) \, dx \, dy$$

7. Evaluate the double integral on the given region.

$$\iint\limits_R \left(2e^y\right) \, dA; R : 0 \le x \le 4 \text{ and } x \le y \le 2x$$

8. Find the volume of the solid bounded above by the graph of $f(x, y) = 8x + 3y + 3$ and below by the rectangle; $R : 1 \le x \le 4 \text{ and } 3 \le y \le 5$.

9. Evaluate the double integral on the given region.

$$\iint\limits_R \left(8x^2y^2\right) \, dA; R : -1 \le x \le 0 \text{ and } \sqrt[3]{x} \le y \le x^3$$

10. Evaluate the double integral.

$$\int_{-1}^2 \int_{-x}^x \left(3x^2y^2\right) \, dy \, dx$$

Answer Key

0.1 Exercises

1. Integer, Rational Number, Real Number

3. Irrational Number, Real Number

5. $\dfrac{-1}{6} > \dfrac{-2}{3}$ and $\dfrac{-1}{6} \geq \dfrac{-2}{3}$

7. $-11 \leq 2$

9. $(-11, -6]$

0.2 Exercises

1. $7x^5y^4, -(x-y), -5xz$

3. Commutative Property of Multiplication

5. Additive Cancellation Property, $\dfrac{-1}{8}y^6 + z$

7. $-25 - 9\pi$

9. $[4, 7]$

0.3 Exercises

1. 64

3. -125

5. $\dfrac{1}{x^5}$

7. $\dfrac{z^2}{1024y^7}$

9. $-64x^6y^6$

0.4 Exercises

1. Real Number, -5

3. $-4y^4z^2$

5. $|z^5|\sqrt{3x}$

7. $\dfrac{x + \sqrt{6x}}{x - 6}$

9. $\dfrac{\sqrt{30zx}}{5x}$

0.5 Exercises

1. y^4

3. $\dfrac{1}{256}$

5. $6z\sqrt[5]{2z}$

7. $10y\sqrt{2y}$

9. $z^{\frac{8}{7}}$

0.6 Exercises

1. $x^2 + 9x + 14$

3. $x^3 - 3x^2 - 3x - 4$

5. $x^2 + 11x$

7. $2y\left(3x - 7y^2\right)$

9. $(z - 6)(z - 7)$

1.1 Exercises

1. One Solution, $y = 4$

3. One Solution, $u = -2$

5. All Real Numbers

7. One Solution, $t = 4$

9. One Solution, $x = \dfrac{17}{26}$

1.2 Exercises

1. $\dfrac{C}{2\pi}$

3. $\dfrac{2A}{h} - B$

5. $\dfrac{A}{2(l+h)} - \dfrac{hl}{(l+h)}$

7. 5 hours

9. $255

1.3 Exercises

1. **a.** $(-\infty, 3)$

   b. ←————————)————→

3. **a.** $[6, \infty)$

   b. ←————————[————→

5. **a.** $[-9, 2]$

   b. ←———[———————]———→

7. **a.** $[-4, \infty)$

   b. $(-\infty, 9)$

   c. $(-\infty, \infty)$

   d. ←————————————→

9. **a.** $[-2, 0]$

   b. ←————————[-]————→

1.4 Exercises

1. $\dfrac{11}{4}, -\dfrac{11}{4}$

3. $0, -3$

5. $-3, -9$

7. $-4, -18$

9. $\dfrac{7}{3}, 0$

1.5 Exercises

1. 3 or -3 or $2\sqrt{2}$ or $-2\sqrt{2}$

3. $-1, -343$

5. $2i$ or $-2i$ or 1 or -1

7. $0, \dfrac{7}{5}, -7$

9. $0, \dfrac{2}{5}, -6$

1.6 Exercises

1. **a.** $x \neq -8$

   b. $x = 40$

3. **a.** $x \neq 4, -8$

   b. $x = -2, -4$

5. **a.** $x \neq -2, -1$

   b. $x = -4$

7. $z = \dfrac{9}{5}, -3$

9. $x = 25$

2.1 Exercises

1.

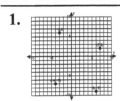

3.

x	$\dfrac{16}{9}$	0	2	$\dfrac{34}{9}$
y	0	$\dfrac{-8}{3}$	$\dfrac{1}{3}$	3

5. **a.** Right Triangle

   b. 28

7.

x	0	9	$4\sqrt{5}$	$-\sqrt{77}$	$\sqrt{65}$
y	9	0	1	2	4

9. $\sqrt{10} + \sqrt{197} + 5\sqrt{5}$

2.2 Exercises

1. Linear and Standard form: $9x + 23y = 0$

3. Linear and Standard form: $-9x + 14y = -37$

5. Not Linear

7. **a.** x-intercept $= \left(\dfrac{1}{2}, 0\right)$ and y-intercept $= (0, 3)$

b.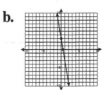

9. **a.** x-intercept $= (-2, 0)$ and y-intercept $=$ absent

b.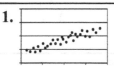

2.3 Exercises

1. Slope $= \dfrac{2}{7}$

3. Slope $= 0$

5. **a.** $y = -4x + 6$

    b. $y = -2$

    c. $y = -6$

7. Slope-intercept form : $y = -x - 1$

9. Standard form : $x = 10$

2.4 Exercises

1. **a.** $y = -4x + \dfrac{9}{2}$

    b. $y = -4x - 18$

3. **a.** $y = \dfrac{-1}{4}x - \dfrac{3}{4}$

    b. $y = 4x + 26$

5. **a.** $y = \dfrac{5}{3}x + \dfrac{1}{3}$

    b. $y = \dfrac{-3}{5}x - \dfrac{62}{5}$

7. Parallel

9. Perpendicular

2.5 Exercises

1.

3. $r = 0.684$, No

5. $\hat{y} = 27.774 - 0.399x$, Yes

7. 0.576, Yes

9. -172.28

3.1 Exercises

1. **a.** $f(8) = 509$

    b. $f(-5) = 197$

    c. $f(x + h) = 8x^2 + 16hx + 8h^2 - 3$

3. **a.** $f(10) = 98$

    b. $f(9) = 79$

5. $\mathbb{R}$

7. $x \neq 9, -5$

9. $x \neq \pm 2$

3.2 Exercises

1. $x = 2$

3. **a.** $R(x) = 20x$

    b. $C(x) = 0.14x + 2780.40$

    c. $P(x) = 19.86x - 2780.40$

    d. $x = 140$

5. $x = 21, 35$

7. $C(x) = \begin{cases} 10.75 & \text{if } 0 < x \le 5 \\ 10.75 + 0.52(x - 5) & \text{if } x > 5 \end{cases}$

9. $A(x) = 127x - x^2$

3.3 Exercises

1. **a.** $(4, 4)$

    b. Zero

    c. A: $(2, 8)$, B: $(5, 5)$

    d.

3. **a.** $y = -2x + \dfrac{11}{3}$

    b. $y = -2x - 15$

5. $f(x) = -4x - 5$

7. **a.** $(4, -4)$

    b. Two: $(6, 0), (2, 0)$

    c. A: $(1, 5)$, B: $(7, 5)$

    d.

9. **a.** $(-3, -4)$

    b. Two: $(-1, 0), (-5, 0)$

    c. A: $(-6, 5)$, B: $(-2, -3)$

    d.

3.4 Exercises

1. $\dfrac{97}{2}, \dfrac{97}{2}$

3. $200, \dfrac{400}{3}$

5. 300 apartments

7. 80000 dollars

9. Maximum, Maximum occurs at $x = -3$;
Maximum is $y = 5$

3.5 Exercises

1. **a.**

    b.

    A: $\left(1, \dfrac{3}{2}\right)$ B: $\left(-1, \dfrac{-3}{2}\right)$

3. **a.**

    b.

    A: $\left(-1, \dfrac{1}{2}\right)$ B: $\left(1, \dfrac{-1}{2}\right)$

5. a.

b.

A: $\left(-1, \dfrac{-1}{5}\right)$ B: $\left(1, \dfrac{1}{5}\right)$

7. a.

b.

A: $\left(-1, \dfrac{-5}{9}\right)$ B: $\left(3, \dfrac{-5}{3}\right)$

9.

3.6 Exercises

1. a. $f(x) = x^3$

b.

Horizontal Shift: Right, Units: 3
Stretch/Compress: None
x-Axis Reflection: No
y-Axis Reflection: No
Vertical Shift: None

c. Domain: $\mathbb{R}$ Range: $\mathbb{R}$

3. a. $f(x) = \dfrac{1}{x}$

b.

Horizontal Shift: Right, Units: 5
Stretch/Compress: None
x-Axis Reflection: Yes
y-Axis Reflection: No
Vertical Shift: None

c. Domain: $(-\infty, 5) \cup (5, \infty)$
Range: $(-\infty, 0) \cup (0, \infty)$

5. a. $f(x) = \sqrt{x}$

b.

Horizontal Shift: Left, Units: 5
Stretch/Compress: Stretch, 2
x-Axis Reflection: Yes
y-Axis Reflection: No
Vertical Shift: Up, Units: 1

c. Domain: $[-5, \infty)$ Range: $(-\infty, 1]$

7. Even

9. y-Axis Symmetry

3.7 Exercises

1. $5 + 2\sqrt{3}i, 5 - 2\sqrt{3}i$

3.

x-intercepts: $(0,0), (-2,0), (-3,0)$
y-intercept: $(0,0)$

5.

x-intercepts: $(-6,0), (-1,0), (2,0), (1,0)$
y-intercept: $(0,-12)$

7. $(-\infty, -5] \cup [0,2]$

9. $(-\infty, -5] \cup [4, \infty)$

3.8 Exercises

1. $y = 0$

3. $y = -1$

5. $x = \dfrac{-7}{2}$

7. a.

b.

c.

9. a.

b.

c.

3.9 Exercises

1. $(-6, -5) \cup (1, \infty)$

3. $(-6, 0)$

5. $(-\infty, -3) \cup (3, \infty)$

7. $(-\infty, -3) \cup (2, 5]$

9. a. $-4, 5$

b. $(-\infty, -4) \cup (5, \infty)$

4.1 Exercises

1.

3.

5.

7.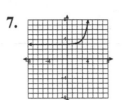

9. $\dfrac{4}{3}$

4.2 Exercises

1. a. $P(t) = 411(1.055)^t$

   b. 50,882 bacteria

3. 236828 people

5. 0.73%

7. $36016.78

9. a. Account 2

   b. $8330.73

4.3 Exercises

1. $3 = \log_c(3.1)$

3.

5. -2

7. 0

9. 16

4.4 Exercises

1. $\dfrac{x^3}{y^2}$

3. x^6

5. $\log(1.7) + 11 + 9\log(x)$

7. $\ln(15) + 9\ln(x) - 6\ln(y)$

9. 11.3

5.1 Exercises

1. $3562.44

3. 25%

5. $425

7. $8.24

9. 40%

5.2 Exercises

1. $48,926.39

3. $18,000

5. $98,497.03

7. 520%

9. $1170

5.3 Exercises

1. $20,148.43

3. $989,433.88

5. $445.13

7. $59,544.96; $15,455.04

9. $45,383.76; $19,616.24

5.4 Exercises

1. $169,146.21

3. 17 months

5. $4235.80

7. $2978.88

9. $91.20

6.1 Exercises

1. Only One Solution, $(10, -31)$

3. Only One Solution, $(-2, -4)$

5. Infinitely Many Solutions

7. No Solution

9. pounds of the alloy containing 23% aluminum $= 39.72$, pounds of the alloy containing 52% aluminum $= 8.28$

6.2 Exercises

1. $\begin{bmatrix} 0 & -35 & -25 \\ 1 & 9 & 6 \end{bmatrix}$

3. **a.** 2×3

   b. -2

   c. -14

5. $\begin{bmatrix} \dfrac{2}{7} & -\dfrac{7}{6} & -\dfrac{1}{6} & \dfrac{5}{7} \\ 1 & -8 & -21 & 5 \\ 8 & 5 & -2 & -7 \end{bmatrix}$

7. No Solution

9. No Solution

6.3 Exercises

1. 0

3. -2

5. 112

7. One Solution, $(-8, -1)$

9. One Solution, $(-6, -1, 1)$

6.4 Exercises

1. $\begin{bmatrix} 1 & 10 \\ 18 & -12 \\ -6 & -9 \end{bmatrix}$

3. Not Possible

5. $r = 8, s = 3$

7. $\begin{bmatrix} 72 & 64 & 16 \\ -36 & -32 & -8 \\ -81 & -72 & -18 \end{bmatrix}$

9. $\begin{bmatrix} 228 & 189 \end{bmatrix}$

6.5 Exercises

1. $\begin{bmatrix} 4 & -7 \\ 2 & 8 \end{bmatrix} \begin{bmatrix} x \\ y \end{bmatrix} = \begin{bmatrix} 2 \\ -10 \end{bmatrix}$

3. $\begin{bmatrix} \dfrac{-1}{7} & \dfrac{-3}{7} \\ \dfrac{-1}{14} & \dfrac{2}{7} \end{bmatrix}$

5. $\begin{bmatrix} -49 & -7 & -10 \\ -8 & -1 & -2 \\ -6 & -1 & -1 \end{bmatrix}$

7. Infinitely Many Solutions, $\left\{ \left(-\dfrac{3}{2}y - \dfrac{1}{2}, y \right) \middle| y \in \mathbb{R} \right\}$

9. One Solution, $x = -5, y = -1, z = 2$

6.6 Exercises

1. $\begin{bmatrix} 1900 \\ 400 \end{bmatrix}$

3. $1800 natural gas, $2400 coal

5. $\begin{bmatrix} 360 \\ 80 \\ 100 \end{bmatrix}$

7. electric

7.1 Exercises

1.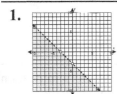

3.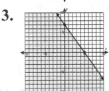

5. a.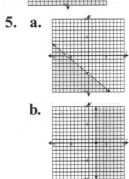

b.

c.

7. a. $x > \dfrac{-10}{7}$ and $x < \dfrac{8}{7}$

b.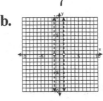

9. a.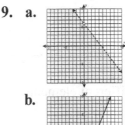

b.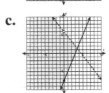

c.

7.2 Exercises

1. Minimum Value of 0 at $(0, 0)$
Maximum Value of 16 at $(2, 0)$

3. Minimum Value of 0 at $(0, 0)$
Maximum Value of 36.88 at $(0.44, 4)$

5. Minimum Value of 0 at $(0, 0)$
Maximum Value of 72 at $(0, 8)$

7. Maximum revenue of $\$34965$ by planting 270 apple trees and 45 peach trees.

9. Minimum value of 7 at $(0, 7)$,
Maximum value does not exist

7.3 Exercises

1. $f(x_1, x_2) = 5x_1 + 6x_2$ attains a maximum value of 38 at $(x_1, x_2) = (4, 3)$.

3. f attains a maximum value of 16 for infinitely many points. For all (x_1, x_2) that lies on the constraint line $x_1 + 2x_2 = 4$, the value of $f(x_1, x_2)$ is equal to $4(x_1 + 2x_2) = 4(4) = 16$.

5. $f(x_1, x_2, x_3) = 5x_1 + 6x_2 + 3x_3$ attains a maximum value of 48 at $(x_1, x_2, x_3) = (3, 2, 7)$.

7.4 Exercises

1. $f(x_1, x_2) = 3x_1 + x_2$ attains a minimum value of 5 at $(x_1, x_2) = (0, 5)$.

3. $f(x_1, x_2, x_3) = x_1 + 2x_2 + x_3$ attains a minimum value of $\frac{17}{3}$ at $(x_1, x_2, x_3) = \left(5, \frac{1}{3}, 0\right)$.

5. $f(x_1, x_2, x_3) = 7x_1 + 10x_2 + 7x_3$ attains a minimum value of 31 at $(x_1, x_2, x_3) = \left(\frac{3}{8}, \frac{15}{8}, \frac{11}{8}\right)$.

7.5 Exercises

1. $f(x_1, x_2) = 4x_1 + 6x_2$ attains a minimum value of 4 at $(x_1, x_2) = (1, 0)$.

3. $f(x_1, x_2, x_3) = 2x_1 - 2x_2 + 6x_3$ attains a maximum value of 20 at $(x_1, x_2, x_3) = (0, 5, 5)$.

5. $f(x_1, x_2, x_3) = 7x_1 + 2x_2 + 3x_3$ attains a minimum value of 20 at $(x_1, x_2, x_3) = (1, 2, 3)$.

8.1 Exercises

1. $|W| = 4$

3. No, because the sets contain different elements.

5. $G = \{x | x \in \mathbb{R}, x > 10\}$

7. Equivalent Sets

9. $A' = \{b, c, f, g, j, k, l, m, o, p, r, t, w, x, y, z\}$

8.2 Exercises

1. $A \cup B = \{1, 2, 3, 4, 5, 6, 7, 8, 10, 12, 14\}$

3. $P \cup (Q \cap R) = \{s, a, m, p, l, e, i\}$

5. Enzo, Marc

7. Cho, Dave, Ella, Hope, Jose, Kaia, Lola, Mia, Sean, Tori, Zack

9. Intersection

8.3 Exercises

1. {CBW, CBH, CIW, CIH, COW, COH, CSW, CSH, JBW, JBH, JIW, JIH, JOW, JOH, JSW, JSH}

3. 506 of the 1500 female students in Lucia's freshman class are from her home state. Aaron would like to determine the chances of her roommate being from her home state.

5. 0

7. $\frac{12}{37}$ Or 0.324324

9. {RR, RB, RY, RG, BR, BB, BY, BG, YR, YB, YY, YG, GR, GB, GY, GG}

8.4 Exercises

1. 5040

3. 42

5. 280

7. 3432

9. 5040

8.5 Exercises

1. $\frac{1}{6545} \approx 0.000153$

3. $\frac{10}{231} \approx 0.043290$

5. $\frac{1}{820} \approx 0.001220$

7. $\frac{1}{15120} \approx 0.000066$

9. $\frac{5}{6} \approx 0.833333$

8.6 Exercises

1. $\dfrac{1}{104} \approx 0.009615$

3. $\dfrac{12}{19} \approx 0.631579$

5. $\dfrac{7}{13} \approx 0.538462$

7. $\dfrac{1}{11} \approx 0.090909$

9. $\dfrac{59}{84}$

8.7 Exercises

1. 1:4

3. 1.148 points

5. 0.6 times per month

7. Expected value $= -\$0.25$

9. Lose, $294.25

9.1 Exercises

1. Cluster Sample

3. Population: U.S. homeowners; Sample: 2617 homeowners polled; Sample Statistic: $18,500

5. Stratified

7. Random Sample

9. 2617 homeowners polled

9.2 Exercises

1.

Grade	Frequency
A	2
B	1
C	8
D	7
F	3

3. 13.9

5. 31.9%

7. 4

9. The graph is not misleading.

9.3 Exercises

1. 22

3. 8.88

5. 208.25

7. Mode because the data have no measurable values.

9. below 47 and above 143

9.4 Exercises

1. 0.0006

3. 0.2066

5. 0.0033

7. 0.2461

9. 0.1563

9.5 Exercises

1. 0.21

3. 0.84

5. 6.36

7. 65.29%

9. 92.23%

9.6 Exercises

1. Area to the right of 80.5

3. Area to the left of 19.5

5. Area between 18.5 and 19.5

7. 0.3974

9. 0.1949

10.1 Exercises

1.

a.

x	y
-1	-7.937
-0.1	-79.937
-0.01	-799.937
-0.001	-7999.937

b. $-\infty$

3. $-\infty$

5. **a.** $\lim\limits_{x \to -6^-} f(x) = -30$

b. $\lim\limits_{x \to -6^+} f(x) = -260$

10.2 Exercises

1. **a.** 3

b. 4

c. Does Not Exist

3. **a.** $-\infty$ or Does Not Exist

b. -1

c. Does Not Exist

10.3 Exercises

1. $\dfrac{-1}{6}$

3. 15

5. ∞

7. $+\infty$

9. 14

10.4 Exercises

1. **a.** Does Not Exist

b. -5

c. -5

d. No

3. 60

10.5 Exercises

1. $\dfrac{-2}{3}$

3. -2

5. -7

7. 5.75

9. -0.03

10.6 Exercises

1. Additional cost for increasing production by one table when x tables are being produced.

3. **a.** The value of the function evaluated at $x = 7$ is -86.

b. The instantaneous rate of change at $x = 7$ is 16.

5. a. $f(19) = 3.928$ The expected number of births for an average woman born in 1989 is $f(19)$.

b. $f'(19) = -0.068$ The rate of change in expected number of births for an average woman born in 1989 is $f'(19)$.

7. 0.13

9. 0.38

10.7 Exercises

1. $y' = -5$

3. $f'(x) = 28x^3$

5. $f'(x) = \dfrac{8}{3}x^{-\frac{2}{3}}$

7. $y' = 1.2x^{-0.8}$

9. $f'(x) = -\dfrac{16}{7}x^{-\frac{5}{7}}$

10.8 Exercises

1. $f'(x) = -20x^4 + 6x^2$

3. $g'(x) = \dfrac{4}{5}x^3 + 25x^4 + 16x$

5. $f'(x) = 16x + 26$

7. 477 people per year

10.9 Exercises

1. $\overline{C}(x) = \dfrac{830}{x} + 2$

3. $D(x) = 42$

5. $R(x) = 20.9x - 1.6x^2$

7. $\overline{C}'(x) = -\dfrac{2500}{x^2} + 0.4$

9. $P'(x) = -0.01x + 2.39$

11.1 Exercises

1. $f'(x) = 9x^2$

3. $h'(4) = 224$

5. $h'(4) = \dfrac{49}{3}$

7. $f'(x) = \dfrac{-8x^4 - 12x^2 - 16x}{\left(8x^2 + 4\right)^2}$

9. $y = 8x - 25$

11.2 Exercises

1. $3\left(-x^3 + 4x + 5\right)\left(5x^4 - 2x + 3\right)^2\left(20x^3 - 2\right) + \left(5x^4 - 2x + 3\right)^3\left(-3x^2 + 4\right)$

3. $-3\left(\dfrac{7x^3 - 9x + 5}{x^3 - 5x + 10}\right)^{-4}\left(\dfrac{\left(x^3 - 5x + 10\right)\left(21x^2 - 9\right) - \left(7x^3 - 9x + 5\right)\left(3x^2 - 5\right)}{\left(x^3 - 5x + 10\right)^2}\right)$

5. $3\left(-2x^3 + 9\right)^2\left(7x^2 + 7\right)^2\left(\left(-2x^3 + 9\right)(14x) + \left(7x^2 + 7\right)\left(-6x^2\right)\right)$

7. $-\dfrac{21}{2}t^2\left(t^3 - 4\right)^{-\frac{3}{2}}$

9. 218 units per day

11.3 Exercises

1. $\dfrac{2x}{9y^2}$

3. $\dfrac{-2y}{3x}$

5. a. $\dfrac{x}{y}$

b. 2

7. $\dfrac{-81}{4}$ units of capital per week

11.4 Exercises

1. Increasing: $(-\infty, \infty)$
Decreasing: $\varnothing$
3. Increasing: $\varnothing$
Decreasing: $(-\infty, -4)$

5. a. 2
  b. Increasing: $(-\infty, \infty)$, Decreasing: $\varnothing$
7. Increasing: $\varnothing$, Decreasing: $(-\infty, 4)$, $(4, \infty)$

11.5 Exercises

1. a. $x = -7, 5$
  b. Local Maxima: $(-7, 540)$,
  Local Minima: $(5, -324)$

3. a. $x = \dfrac{-9}{2}, \dfrac{9}{2}$
  b. Local Maxima: $\left(\dfrac{-9}{2}, -36\right)$,
  Local Minima: $\left(\dfrac{9}{2}, 36\right)$

5. Increasing on $(0, 56)$, Decreasing on $(56, 90)$

11.6 Exercises

1. Absolute max: $(8, 160)$
Absolute min: $(2, -20)$
3. Absolute max: $(6, 540)$
Absolute min: $(-7, -266)$

5. Absolute max: $(5, 11)$
Absolute min: $(-8, -28)$
7. Absolute max: $(9, -108)$
Absolute min: $(2, -255)$
9. $x = 19$ units

12.1 Exercises

1. $f''(x) = -4 + \dfrac{10}{9} x^{\frac{-5}{3}}$
3. a. Concave Up: $(1, \infty)$, Concave Down: $(-\infty, 1)$
  b. Points of Inflection: $(1, -20)$

5. a. Concave Up: $(-2, 2)$,
  Concave Down: $(-\infty, -2)$, $(2, \infty)$
  b. Points of Inflection: $\left(-2, \sqrt[4]{60}\right)$, $\left(2, \sqrt[4]{60}\right)$

12.2 Exercises

1. Local Maxima: $(9, 7)$,
Local Minima: No Local Minima
3. a. $f''(x) = 2 + x^{-\frac{3}{2}}$
  b. Local Maxima: No Local Maxima,
  Local Minima: $(1, -1)$

5. Local Maxima: $(1, -25)$, Local Minima: $(2, -26)$
7. a. $f''(x) = 12x^2 - 36$
  b. Local Maxima: $(0, 81)$,
  Local Minima: $(3, 0)$, $(-3, 0)$

12.3 Exercises

1.

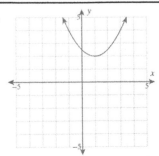

3. a. $f'(x) = -2x^3 + 6x^2$, $f''(x) = -6x^2 + 12x$
  b. Increasing on $(-\infty, 3)$, Decreasing on $(3, \infty)$
  c. Concave up on $(0, 2)$,
  Concave down on $(-\infty, 0)$, $(2, \infty)$
  d. No local minima, Local maximum at $x = 3$,
  Inflection points at $x = 0, 2$

12.4 Exercises

1. a. $x = \dfrac{4}{3}, -4$

    b. $y = 0$

    c. None

3. a. Never increasing,
    Decreasing on $(-\infty, 2), (2, \infty)$

    b. Concave up on $(2, \infty)$,
    Concave down on $(-\infty, 2)$

    c. Local minima: None, Local maxima: None,
    Inflection point(s): None

    d. Vertical asymptote(s): $x = 2$, Horizontal
    asymptote(s): $y = 0$,
    Oblique asymptote(s): None

12.5 Exercises

1. 80 wind chimes per order, 5 orders per year

3. 30 tennis rackets per order, 4 orders per year

5. $1540 per diamond bracelet.

12.6 Exercises

1. 24 ft by 24 ft by 64 ft

3. 11.00 miles

5. a. The maximum height is 8000 feet

    b. 47 seconds

    c. 400 ft / sec

7. The minimum surface area is 1452 in.^{2}

13.1 Exercises

1. Absolute Minimum: $(2, -4.64)$;
Absolute Maximum: $(4.5, 24.65)$

3. a. $N(8) = 320$

    b. 6

5. $\dfrac{4 - 3\ln\left(x^4\right)}{x^4}$

7. $\dfrac{34 - 69x}{68x^6 - 92x^7} - \dfrac{5\ln\left(17x^2 - 23x^3\right)}{4x^6}$

9. $\dfrac{32x}{9 + 8x^2} - \dfrac{28x}{7x^2 + 4}$

13.2 Exercises

1. a. $-4x^2e^x - 8xe^x$

    b. $-4x^2e^x - 16xe^x - 8e^x$

3. $\dfrac{2}{3}e^{-27} - 80e^{-27}\ln(6)$

5. a. $-1.6xe^{-0.8x^2+1}$

    b. Decreasing Intervals: $(0, \infty)$,
    Increasing Intervals: $(-\infty, 0)$

7. $\left(-\dfrac{\sqrt{30}}{4}, -4.583\right), \left(\dfrac{\sqrt{30}}{4}, 4.583\right)$

13.3 Exercises

1. a. $f(t) = 2900e^{0.033t}$

    b. 3421

    c. 2022

3. 9.5 years

5. 19.3 years

7. 40.6 minutes

13.4 Exercises

1. a. $\dfrac{60 - x}{x}$

b. 30

3. a. $\dfrac{12 - 2\sqrt{x}}{\sqrt{x}}$

b. 16

5. 190

7. $x < 19$

13.5 Exercises

1. 21

3. 0

5. ∞^0

7. e^{-15}

13.6 Exercises

1. $du = (18t + 48)dt$

3. $\Delta y - dy = -5.612$

5. $4 + \dfrac{3}{8}$

7. $5 + \dfrac{13}{100}$

9. $63.04

14.1 Exercises

1. $-x^7 - 5x + C$

3. $-\dfrac{16}{9}x^{\frac{1}{2}} - 5x + C$

5. $-\dfrac{21}{4}x^{\frac{4}{3}} - 6x^{\frac{3}{2}} + C$

7. $\dfrac{4}{3}t^{\frac{3}{4}} + \dfrac{3}{7}t^{-7} + C$

9. $3x - 5\ln(|x|) + \dfrac{2}{5}x^{-5} + C$

14.2 Exercises

1. $\dfrac{1}{8}(x + 2)^8 + C$

3. $e^{2x^9 + 9} + C$

5. $(9t + 7)^{\frac{2}{3}} + C$

7. $-6\ln\big(|e^t + 7|\big) + C$

9. $-\dfrac{5}{18}\big(4y^3 - 1\big)^{\frac{3}{2}} + C$

14.3 Exercises

1. Area $= 24$

3. 405.00

5. 1.58

7. 13.84

14.4 Exercises

1. $\dfrac{-65}{4}$

3. $\dfrac{243}{2}$

5. $9 + 4e^4 - 4e \approx 216.52$

7. 128

9. $\dfrac{63135}{8} = 7891.88$

14.5 Exercises

1. 495

3. $20 + 20\big(e^2 - e^{0.4}\big)$

5. $2624.46

7. $13693.91

9. 593 wild animals

14.6 Exercises

1. $2250

3. **a.** $(25, 191)$

    b. $250.00

    c. $937.50

5. $\dfrac{38}{3} + \dfrac{1}{0.4}e^{-1.2} - \dfrac{1}{0.4}e^{-0.4} \approx 11.74$

7. $11 + e^{-2} \approx 11.14$

14.7 Exercises

1. $y = ke^{x^4}$

3. $y = Ce^{-4x}$

5. $y = \dfrac{2}{1 + Ce^{-8x}}$

15.1 Exercises

1. $\dfrac{\ln(x)\, x^5}{5} - \dfrac{x^5}{25} + C$

3. $\dfrac{\ln(4x)\, x^8}{8} - \dfrac{x^8}{64} + C$

5. $\dfrac{-x(x-9)^{-3}}{3} - \dfrac{(x-9)^{-2}}{6} + C$

7. $\dfrac{-9xe^{2x}}{2} + \dfrac{9e^{2x}}{4} + C$

9. $\dfrac{36}{5}$

15.2 Exercises

1. $17,049.37

3. $2,156,260.41

5. $113,447.64

7. $2825.96

9. $3934.69

15.3 Exercises

1. $9\left(\dfrac{x^6 \ln(x)}{6} - \dfrac{x^6}{36}\right) + C$

3. $\ln\left(\left|x + \sqrt{x^2 - 144}\right|\right) + C$

5. $\dfrac{1}{4}\ln\left(\left|\dfrac{x+5}{x+9}\right|\right) + C$

7. $-\dfrac{5}{17}\ln\left(\left|\dfrac{x}{2x-17}\right|\right) + C$

9. $e^{-2x}\left(\dfrac{-x^2}{2} - \dfrac{x}{2} - \dfrac{1}{4}\right) + C$

15.4 Exercises

1. T_8: 6724

3. T_8: 9.7738

5. T_{10}: 20.0162

7. $\dfrac{130}{3}$ cubic feet

15.5 Exercises

1. 0

3. convergent, $\dfrac{3}{2e^8}$

5. divergent

7. convergent, $\dfrac{1}{e^4}$

9. convergent, $\dfrac{2}{e^9}$

15.6 Exercises

1. 654π

3. 130π

5. $\dfrac{\pi e^{20} - \pi e^4}{4}$

7. $\dfrac{40\pi}{7}$

9. $9\pi \ln\left(\dfrac{3}{2}\right)$

16.1 Exercises

1. 108

3. $\dfrac{-33\sqrt{10}}{5}$

5. **a.** 8040 units

   b. 16079 units

7. **a.** $C(x,y) = 5000 + 3.20x + 3.50y$

   b. \$5807.00

16.2 Exercises

1. **a.** $\dfrac{xy^3}{\sqrt{16+x^2}}$

   b. $3y^2\sqrt{16+x^2}$

3. **a.** $\dfrac{-20x^4 y^2}{-4x^5 + 5y^3}$

   b. $\dfrac{15y^4}{-4x^5 + 5y^3} + 2y\ln\left(-4x^5 + 5y^3\right)$

16.3 Exercises

1. Local Maxima: None, Local Minima: None, and Saddle Points: $\left(\dfrac{1}{3}, \dfrac{-1}{2}, \dfrac{-103}{12}\right)$

3. Local Maxima: None, Local Minima: $(2,1,-17)$, and Saddle Points: $(-2,1,47)$

5. Local Maxima: $(2,-3,48)$, Local Minima: None, and Saddle Points: $(2,1,16)$

7. Local Maxima: $(-2,1,-12)$, Local Minima: None, and Saddle Points: None

16.4 Exercises

1. $f(-10,-8) = 84$

3. 30 units in Phoenix, 198 units in Charleston

5. \$823,200

16.5 Exercises

1. **a.** $y = 1.10x + 1.80$

   b.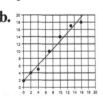

3. $y = -0.25x + 15.96$

5. **a.** $y = 1.30x + 42.20$

   b. \$51.30

16.6 Exercises

1. 20

3. $28\ln(2)$

5. $\dfrac{e^{22} - e^{10} - e^{20} + e^8}{6}$

7. $e^8 - 2e^4 + 1$

9. $\dfrac{4}{9}$

Notes

Notes

Notes

Notes

Notes

Notes

Notes

Notes

Notes

Notes